T0207016

Quaternions for Computer Graphics

John Vince

Quaternions for Computer Graphics

Second Edition

John Vince
Bournemouth University
Poole, UK

ISBN 978-1-4471-7511-7 ISBN 978-1-4471-7509-4 (eBook)
https://doi.org/10.1007/978-1-4471-7509-4

This Springer imprint is published by the registered company Springer-Verlag London Ltd. part of Springer Nature.
The registered company address is: The Campus, 4 Crinan Street, London, N1 9XW, United Kingdom

This book is dedicated to my wife Heidi.

Preface

When I was studying to be an electrical engineer, I came across complex numbers, which were used to represent out-of-phase voltages and currents using the j operator. I believe that the letter j was used, rather than i, because the latter stood for electrical current. So from the very start of my studies, I had a clear mental picture of the imaginary unit as a rotational operator which could advance or retard electrical quantities in time.

When events dictated that I would pursue a career in computer programming—rather than electrical engineering—I had no need for complex numbers, until Mandlebrot's work on fractals emerged. But that was a temporary phase, and I never needed to employ complex numbers in any of my computer graphics software. However in 1986, when I joined the flight simulation industry, I came across an internal report on quaternions, which were being used to control the rotational orientation of a simulated aircraft.

I can still remember being completely bemused by quaternions, simply because they involved so many imaginary terms. However, after much research, I started to understand what they were, but not how they worked. Simultaneously, I was becoming interested in the philosophical side of mathematics and trying to come to terms with the 'real meaning' of mathematics through the writing of Bertrand Russell. Consequently, concepts such as i were an intellectual challenge.

I am now comfortable with the idea that imaginary i is nothing more than a symbol, and in the context of algebra permits $i^2 = -1$ to be defined. And I believe it is futile trying to discover any deeper meaning to its existence. Nevertheless, it is an amazing object within mathematics, and I often wonder whether there could be similar objects waiting to be invented.

When I started writing books on mathematics for computer graphics, I studied complex analysis in order to write with some confidence about complex quantities. It was then that I discovered the historical events behind the invention of vectors and quaternions, mainly through Michael Crowe's excellent book: *A History of Vector Analysis*. This book brought home to me the importance of understanding how and why mathematical invention takes place.

I then came across Simon Altmann's book: *Rotations, Quaternions, and Double Groups* which provided further information concerning the demise of quaternions in the nineteenth century. Altmann is very passionate about securing recognition for the mathematical work of Olinde Rodrigues, who published a formula that is very similar to that generated by Hamilton's quaternions. The important aspect of Rodrigues' publication was that it was made three years before Hamilton's invention of quaternions in 1843. However, Rodrigues did not invent quaternion algebra—that prize must go to Hamilton—but he did understand the importance of half-angles in the trigonometric functions used to rotate points about an arbitrary axis.

Anyone who has used the Euler transforms will be aware of their shortcomings, especially their Achilles' heel: gimbal lock. Therefore, any device that can rotate points about an arbitrary axis is a welcome addition to a programmer's toolkit. There are many techniques for rotating points and frames in the plane and space, which I covered in some detail in my book: *Rotation Transforms for Computer Graphics*. That book also covers the Euler-Rodrigues parameterisation and quaternions, but it was only after submitting the manuscript for publication, that I decided to write this book dedicated to quaternions and how and why they were invented and their application to computer graphics.

Whilst researching this book, it was extremely instructive to read some of the early books and papers by William Rowan Hamilton and his friend P. G. Tait. I now understand how difficult it must have been to fully comprehend the significance of quaternion, and how they could be harnessed. At the time, there was no major demand to rotate points about an arbitrary axis; however, a mathematical system was required to handle vectorial quantities. In the end, quaternions were not the flavour of the month and slowly faded from the scene. Nevertheless, the ability to represent vectors and manipulate them arithmetically was a major achievement for Hamilton, even though it was the foresight of Josiah Gibbs to create a simple and workable algebraic framework.

I hope that this second edition improves the original book as intended. I have reviewed all the texts and algebraic notations and redesigned all the figures to include colour. I have also adjusted the bibliography to be localised for each chapter and enlarged the index. Specifically, I have introduced a chapter describing how triples led to the invention of quaternions. Finally, I have expanded the reference to Olinde Rodrigues and the section on interpolation.

There are different ways to represent a quaternion, but the one I like the best is an ordered pair, which I discovered in Simon Altmann's book. In no way would I consider myself an authority on quaternions. I simply want to communicate how I understand them, which hopefully will be useful for you.

This book now divides into nine chapters. The first and last chapters introduce and conclude the book, with seven chapters covering the following subjects. Chapter 2 is on number sets and algebra and reviews the notation and language relevant to the rest of the book. There are sections on number sets, axioms, ordered pairs, groups, rings and fields. This prepares the reader for the non-commutative quaternion product, and why quaternions are described as a division ring.

Chapter 3 reviews complex numbers and shows how they can be represented as an ordered pair and a matrix. Chapter 4 continues this theme by introducing the complex plane and showing the rotational features of complex numbers. It also prepares the reader for the question that was asked in the early nineteenth century: could there be a 3-D equivalent of a complex number? Chap. 5 traces the emergence of quaternions through the intermediate stage of triples and shows how Hamilton introduced a third imaginary k and wrestled with its integration with i and j.

Chapter 6 answers this question by describing Hamilton's invention: quaternions and the associated algebra. I have included some historical information so that the reader appreciates the significance of Hamilton's work. Although ordered pairs are the main form of notation, I have also included matrix notation.

To prepare the reader for the rotational qualities of quaternions, Chap. 7 reviews 3-D rotation transforms, especially Euler angles, and gimbal lock. I also develop a matrix for rotating a point about an arbitrary axis using vectors and matrix transforms.

Chapter 8 is the focal point of the book and describes how quaternions rotate vectors about an arbitrary axis. The chapter begins with some historical information and explains how different quaternion products rotate points. Although quaternions are readily implemented using their complex form or ordered-pair notation, they also have a matrix form, which is developed from the first principles. The chapter continues with sections on rotating about an offset-axis, rotating frames of reference, interpolating quaternions, and converting between quaternions and a rotation matrix. Each chapter contains many practical examples to show how equations are evaluated, and where relevant, further worked examples are shown at the end of the chapter.

Preparing this second edition has been a very enjoyable experience, and I trust that you will also enjoy reading it and discover something new from its pages.

I would like to thank Dr. Tony Crilly, Reader Emeritus at Middlesex University, for reading the original manuscript and correcting and clarifying my notation and explanations. I trust implicitly his knowledge of mathematics and I am grateful for his advice and expertise. However, I still take full responsibility for any algebraic *faux pas* I might have made.

I would also like to thank Prof. Patrick Riley, who read some early drafts of the manuscript and posed some interesting technical questions about quaternions. Such questions made me realise that some of my descriptions of quaternions required further clarification, which hopefully have been rectified.

I am not sure whether this is my last book. If it is, I would like to thank Helen Desmond, Editor for Computer Science, Springer UK, for her professional support during the past years. If it is not my last book, then I look forward to working with her again on another project.

Breinton, UK John Vince

Contents

Chapter 1
Introduction

1.1 Rotation Transforms

In computer graphics transforms are used to modify the position and orientation of an object or a virtual camera. Such transforms generally comprise: scale, translation and rotation. The first two transforms are relatively easy to visualise, but rotations can cause problems—due to the need to construct a rotation transform from individual rotations about the x-, y- and z-axes. Although such transforms work, they are far from perfect. What really is required, is a technique that is intuitive, simple and accurate.

Over the years, rotation transforms have embraced direction cosines, Euler angles, Euler-Rodrigues parameterisation, quaternions and multivectors. The last two techniques are the most recent, and are historically related. However, the subject of this book is quaternions, and how they can be used within computer graphics.

1.2 The Reader

This book is aimed at readers studying or working in computer graphics and require an overview of quaternions. They are probably the same people I have encountered asking questions on Internet forums about quaternions, how they work, and how they can be coded. Hopefully, this book will answer most of these questions.

1.3 Aims and Objectives of This Book

The primary aim of this book is to introduce the reader to the subject of quaternions and how they can be used to rotate points about an arbitrary axis. A secondary aim is to make the reader aware of the human dimension behind all mathematical discovery.

J. Vince, *Quaternions for Computer Graphics*,
https://doi.org/10.1007/978-1-4471-7509-4_1

Personally, I believe that we must never lose sight of the fact that mathematicians are human beings; and although they may be endowed with extraordinary mathematical skills, they fall in love, marry, raise families, die, and leave behind an amazing edifice of knowledge, from which we all benefit.

Those readers interested in the human dimension of mathematics are encouraged to read about the invention of quaternions and discover how love, intrigue, tenacity, inspiration and dedication resulted in a major mathematical discovery. One book that is essential reading is Michael Crowe's *A History of Vector Analysis* [1]. This provides a thorough analysis of the events leading up to the invention of quaternions, and the emergence of vector analysis. A second book is Simon Altmann's *Rotations, Quaternions, and Double Groups* [2], which, apart from providing a modern analysis of quaternion algebra, introduces Olinde Rodrigues, who invented some of the mathematics associated with quaternions three years before William Rowan Hamilton, the man usually acknowledged as the father of quaternions.

Simon Altmann's analysis of quaternion algebra has had a profound influence upon my own views about quaternions, and I have tried to communicate this in the following chapters. In particular, I have adopted the idea of using ordered pairs to represent a quaternion.

The over-riding objective of this book is to enable the reader to design and code quaternion algorithms. After reading the book, this should be a trivial exercise, in spite of the fact that we are dealing with a four-dimensional mathematical object.

1.4 Mathematical Techniques

Once the reader has understood quaternions they will be regarded as easy. However, if this is the first time you have encountered them, they could appear strange. But like all things, familiarity brings understanding and confidence.

In order to describe quaternions I need to call upon a little trigonometry, some vector theory and matrix algebra. And as quaternions are described as 'hyper-complex numbers' a chapter is included on complex numbers.

1.5 Assumptions Made in This Book

Very often, people working in computer graphics—like myself—did not have the opportunity to study mathematics to a level often employed in technical literature. Consequently, I have deliberately written the book with a gentle introduction to formal mathematical notation that progresses through complex algebra towards quaternion algebra. However, feel free to skip over these introductory chapters if you are familiar with the subject material.

References

1. Crowe, M.J.: A History of Vector Analysis. Dover, New York (1994)
2. Altmann, S.L.: Rotations, Quaternions and Double Groups. Dover, New York (1986) ISBN-13: 978-0-486-44518-2

Chapter 2
Number Sets and Algebra

2.1 Introduction

In this chapter we review some basic ideas of number sets, and how they are manipulated arithmetically and algebraically. We look briefly at expressions and equations and the rules used for their construction and evaluation. These, in turn, reveal the need to extend every-day numbers with so called complex numbers. The second part of the chapter is used to define groups, rings and fields.

2.2 Number Sets

2.2.1 *Natural Numbers*

Natural numbers are the whole numbers 1, 2, 3, 4, etc., and by definition (DIN 5473), the *set* of natural numbers and zero $\{0, 1, 2, 3, 4, \ldots\}$ are represented by the symbol $\mathbb{N}$ and we express this assignment using:

$$\mathbb{N} = \{0,\ 1,\ 2,\ 3,\ 4,\ \ldots\}.$$

The statement

$$k \in \mathbb{N}$$

implies that k belongs to the set $\mathbb{N}$, where $\in$ means *belongs to*, or in other words, k is a natural number. We employ this notation throughout this book to ensure that there is no confusion about the type of numerical quantity being used.

$\mathbb{N}^*$ is used to represent the set $\{1,\ 2,\ 3,\ 4,\ \ldots\}$, which leaves out zero.

J. Vince, *Quaternions for Computer Graphics*,
https://doi.org/10.1007/978-1-4471-7509-4_2

2.2.2 Real Numbers

Decimal numbers form the set of *reals* identified by $\mathbb{R}$. Such numbers are signed and can be organised as a line which stretches between $\pm$ infinity, and includes zero. The concept of infinity is a strange one, and was investigated by the German mathematician Georg Cantor (1845–1918). Cantor also invented set theory and proved that real numbers are more numerous than natural numbers. Fortunately, we don't need to employ such concepts within this book.

The set of reals $\mathbb{R}$, is employed to denote dimension, where $\mathbb{R}^2$ stands for two-dimensional, $\mathbb{R}^3$ stands for three-dimensional, and $\mathbb{R}^n$ stands for n-dimensional.

2.2.3 Integers

The set of *integers* $\mathbb{Z}$ embrace the natural numbers and their negatives:

$$\mathbb{Z} = \{\dots, -3, \ -2, \ -1, \ 0, \ 1, \ 2, \ 3, \ \dots\}.$$

$\mathbb{Z}$ stands for *Zahlen*—the German for 'numbers'.

2.2.4 Rational Numbers

The set of *rational* numbers is $\mathbb{Q}$, and contains numbers of the form:

$$\frac{a}{b}, \quad a, b \in \mathbb{Z}, \quad b \neq 0.$$

2.3 Arithmetic Operations

We manipulate numbers using the arithmetic operations addition, subtraction, multiplication and division, whose result is *closed* or not, or *undefined*, depending on the underlying set. For example, when we add two natural numbers together, the result is *always* another natural number and therefore, the operation is closed:

$$3 + 4 = 7.$$

However, when we subtract two natural numbers, the result may not necessarily be a natural number. For instance, although

$$6 - 2 = 4$$

is a closed operation,

$$2 - 6 = -4$$

is not closed, because -4 is not a member of the set of natural numbers.

The product of two natural numbers is *always* a closed operation, however, division causes some problems. To begin with, dividing an even natural number by 2 is a closed operation:

$$\frac{16}{2} = 8.$$

Whereas, dividing an odd natural number by an even natural number gives rise to a decimal quantity:

$$\frac{7}{2} = 3.5$$

and does not close because 3.5 does not belong to the set of natural numbers. In the language of sets, this is written

$$3.5 \notin \mathbb{N}$$

where $\notin$ means *does not belong to*.

Multiplying any number by zero results in zero—which is a closed operation; however, dividing any number by zero is undefined, and has to be excluded.

Real numbers do not have any of the problems associated with natural numbers, and there is closure on addition, multiplication and division:

$$\begin{aligned} a + b = c, &\quad a, b, c \in \mathbb{R} \\ ab = c, &\quad a, b, c \in \mathbb{R} \\ \frac{a}{b} = c, &\quad a, b, c \in \mathbb{R} \text{ and } b \neq 0. \end{aligned}$$

2.4 Axioms

When we construct algebraic expressions we employ specific laws called *axioms*. For addition and multiplication, we know that the grouping of numbers makes no difference to the end result: e.g. $2 + (4 + 6) = (2 + 4) + 6$ and $2 \times (3 \times 4) = (2 \times 3) \times 4$. This is the *associative axiom* and is expressed as:

$$\begin{aligned} a + (b + c) &= (a + b) + c \\ a(bc) &= (ab)c. \end{aligned}$$

We also know that order makes no difference to the end result when adding or multiplying: e.g. $2 + 6 = 6 + 2$ and $2 \times 6 = 6 \times 2$. This is the *commutative axiom* and is expressed as:

$$a + b = b + a$$
$$ab = ba.$$

Algebraic expressions contain all sorts of products involving a single real number and a string of reals that obey the *distributive axiom*:

$$a(b + c) = ab + ac$$
$$(a + b)(c + d) = ac + ad + bc + bd.$$

The reason why we have reviewed these axioms is that they should not be regarded as carved in mathematical stone, and apply to everything that is invented. For when we come to quaternions we discover that they do not obey the commutative axiom, which is not that strange. If you have used matrices you will know that matrix multiplication is also non-commutative, but is associative.

2.5 Expressions

Using the above axioms we are able to construct all sorts of expressions such as:

$$a(2 + c) - \frac{d}{e} + a - 10$$
$$\frac{g}{ac - bd} + \frac{h}{de - fg}.$$

We also employ notation for raising a quantity to some power such as n^2. This notation introduces another set of observations:

$$a^n a^m = a^{n+m}$$
$$\frac{a^n}{a^m} = a^{n-m}$$
$$(a^n)^m = a^{nm}$$
$$\frac{a^n}{a^n} = a^0 = 1$$
$$\frac{1}{a^n} = a^{-1}$$
$$a^{\frac{1}{n}} = \sqrt[n]{a}.$$

Next, we have to include all sorts of functions such as square-roots, sines and cosines, which may seem rather innocent. But we must be wary of them. For example, $\sqrt{16} = \pm 4$, whereas, there is no natural or real number solution for $\sqrt{-16}$. Consequently, the expression $\sqrt{a}$ has no real roots if $a < 0$.

Similarly, when working with trigonometric functions such as sine and cosine, we must remember that these take on a range of values between -1 and $+1$, including 0, which means that if they are employed as a denominator, the result could be undefined. For example, this expression is undefined if $\sin\alpha = 0$

$$\frac{a}{\sin\alpha}.$$

2.6 Equations

Next, we come to equations where we assign the value of an expression to a variable. In most situations the assignment is straightforward and leads to a real result such as

$$x^2 - 16 = 0$$

where $x = \pm 4$. But what is interesting is that just by reversing the sign to

$$x^2 + 16 = 0$$

we create an equation for which there is no real solution. However, there is a *complex* solution, which is the subject of Chap. 3.

2.7 Ordered Pairs

An *ordered pair* or *couple* $(a,\ b)$ is an object having two entries, coordinates or projections, where the first or left entry, is distinguishable from the second or right entry. For example, $(a,\ b)$ is distinguishable from $(b,\ a)$ unless $a = b$. Perhaps the best example of an ordered pair is $(x,\ y)$ that represents a point in $\mathbb{R}^2$, where the order of the entries is always the x-coordinate followed by the y-coordinate.

Ordered pairs and ordered triples are widely used in computer graphics to represent points in $\mathbb{R}^2 : (x,\ y)$, points in $\mathbb{R}^3 : (x,\ y,\ z)$, and colour values such as (r, g, b) and (h, s, v). In these examples, the fields are all real values. There is nothing to stop us from developing an algebra using ordered pairs that behaves like another algebra, and we will do this for complex numbers in Chap. 3 and quaternions in Chap. 6. For the moment, let's explore some ways ordered pairs can be manipulated.

Say we choose to describe a generic ordered pair as

$$a = (a_1,\ a_2), \quad a_1, a_2 \in \mathbb{R}.$$

We will define the addition of two such objects as

$$\begin{aligned} a &= (a_1, \ a_2) \\ b &= (b_1, \ b_2) \\ a + b &= (a_1 + b_1, \ a_2 + b_2). \end{aligned}$$

For example:

$$\begin{aligned} a &= (2, \ 3) \\ b &= (4, \ 5) \\ a + b &= (6, \ 8). \end{aligned}$$

We will define the product as

$$ab = (a_1 b_1, \ a_2 b_2)$$

which, using the above values, results in

$$ab = (8, \ 15).$$

Remember, we are in charge, and we define the rules.

Another rule will control how an ordered pair responds to scalar multiplication. For example:

$$\begin{aligned} \lambda(a_1, \ a_2) &= (\lambda a_1, \ \lambda a_2), \quad \lambda \in \mathbb{R} \\ 3(2, \ 3) &= (6, \ 9). \end{aligned}$$

With the above rules, we are in a position to write

$$\begin{aligned} (a_1, \ a_2) &= (a_1, \ 0) + (0, \ a_2) \\ &= a_1(1, \ 0) + a_2(0, \ 1) \end{aligned}$$

and if we square these unit ordered pairs $(1, \ 0)$ and $(0, \ 1)$ using the product rule, we obtain

$$\begin{aligned} (1, \ 0)^2 &= (1, \ 0) \\ (0, \ 1)^2 &= (0, \ 1) \end{aligned}$$

which suggests that they behave like real numbers, and is not unexpected.

This does not appear to be very useful, but wait and see what happens in the context of complex numbers and quaternions.

2.8 Groups, Rings and Fields

Mathematicians employ a bewildering range of names to identify their inventions, which seemingly, appear on a daily basis. Even the name 'quaternion' is not original, and appears throughout history often in the context of 'a quaternion of soldiers':

> The Romans detached a quaternion or four men for a night guard ... [1].

Without becoming too formal, let's explore some more mathematical structures that are relevant to the ideas contained in this book.

2.8.1 Groups

We have already covered the idea of a set, and what it means to belong to a set. We have also discovered that when we apply certain arithmetic operations to members of a set we can secure closure, non-closure, or the result is undefined.

When combining sets with arithmetic operations, it is convenient to create another entity: a *group*, which is a set, together with the axioms describing how elements of the set are combined. The set might contain numbers, matrices, vectors, quaternions, polynomials, etc., and are represented below as a, b and c.

The axioms employ the '$\circ$' symbol to represent any binary operation such as $+, -, \times$. And a group is formed from a set and a binary operation. For example, we may wish to form a group of integers under addition: $(\mathbb{Z}, +)$, or we may wish to examine whether quaternions form a group under the operation of multiplication: $(\mathbb{H}, \times)$.

To be a group, **all** the following axioms **must** hold for the set S. In particular, there must be a special *identity element* $e \in S$, and for each $a \in S$ there must exist an *inverse element* $a^{-1} \in S$, so that the following axioms are satisfied:

$$
\begin{aligned}
\text{Closure:} &\quad a \circ b \in S, & a, b \in S.\\
\text{Associativity:} &\quad (a \circ b) \circ c = a \circ (b \circ c), & a, b, c \in S.\\
\text{Identity:} &\quad a \circ e = e \circ a = a, & a, e \in S.\\
\text{Inverse:} &\quad a \circ a^{-1} = a^{-1} \circ a = e, & a, a^{-1}, e \in S.
\end{aligned}
$$

We describe a group as $(S, \circ)$, where S is the set and '$\circ$' the operation. For instance, $(\mathbb{Z}, +)$ is the group of integers under the operation of addition, and $(\mathbb{R}, \times)$ is the group of reals under the operation of multiplication.

Let's bring these axioms to life with three examples.

$(\mathbb{Z}, +)$: The integers $\mathbb{Z}$ form a group under the operation of addition:

$$\begin{aligned}
\text{Closure:} &\quad -23 + 24 = 1.\\
\text{Associativity:} &\quad (2+3)+4 = 2+(3+4) = 9.\\
\text{Identity:} &\quad 2+0 = 0+2 = 2.\\
\text{Inverse:} &\quad 2+(-2) = (-2)+2 = 0.
\end{aligned}$$

$(\mathbb{Z}, \times)$: The integers $\mathbb{Z}$ do **not** form a group under multiplication:

$$\begin{aligned}
\text{Closure:} &\quad -2 \times 4 = -8.\\
\text{Associativity:} &\quad (2 \times 3) \times 4 = 2 \times (3 \times 4) = 24.\\
\text{Identity:} &\quad 2 \times 1 = 1 \times 2 = 2.\\
\text{Inverse:} &\quad 2^{-1} = 0.5 \quad (0.5 \notin \mathbb{Z}).
\end{aligned}$$

Also, the integer 0 has no inverse.

$(\mathbb{Q}, \times)$: The group of non-zero rational numbers form a group under multiplication:

$$\begin{aligned}
\text{Closure:} &\quad \tfrac{2}{5} \times \tfrac{2}{3} = \tfrac{4}{15}.\\
\text{Associativity:} &\quad \left(\tfrac{2}{5} \times \tfrac{2}{3}\right) \times \tfrac{1}{2} = \tfrac{2}{5} \times \left(\tfrac{2}{3} \times \tfrac{1}{2}\right) = \tfrac{2}{15}.\\
\text{Identity:} &\quad \tfrac{2}{3} \times \tfrac{1}{1} = \tfrac{1}{1} \times \tfrac{2}{3} = \tfrac{2}{3}.\\
\text{Inverse:} &\quad \tfrac{2}{3} \times \tfrac{3}{2} = \tfrac{1}{1} \quad \left(\text{where } \tfrac{3}{2} = \left(\tfrac{2}{3}\right)^{-1}\right).
\end{aligned}$$

2.8.2 *Abelian Group*

Lastly, an *abelian group*, named after the Norwegian mathematician Neils Henrik Abel (1802–1829), is a group where the order of elements does not influence the result. i.e. the group is commutative. Thus there are five axioms: closure, associativity, identity element, inverse element, and commutativity:

$$\text{Commutativity:} \quad a \circ b = b \circ a, \quad a, b \in S.$$

For example, the set of integers forms an abelian group under ordinary addition $(\mathbb{Z}, +)$. However, because 3-D rotations do not generally commute, the set of all rotations in 3-D space forms a non-commutative group.

2.8.3 Rings

A *ring* is an extended group, where we have a set of objects which can be added and multiplied together, subject to some precise axioms. There are rings of real numbers, complex numbers, integers, matrices, equations, polynomials, etc. A ring is formally defined as a system where $(S, +)$ and $(S, \times)$ are abelian groups and the distributive axioms:

$$\begin{aligned}
\text{Additive associativity:} &\quad a+(b+c)=(a+b)+c, && a,b,c\in S.\\
\text{Multiplicative associativity:} &\quad a\times(b\times c)=(a\times b)\times c, && a,b,c\in S.\\
\text{Distributivity:} &\quad a\times(b+c)=(a\times b)+(a\times c), && \text{and}\\
&\quad (a+b)\times c=(a\times c)+(b\times c), && a,b,c\in S.
\end{aligned}$$

For example, we already know that the integers $\mathbb{Z}$ form a group under the operation of addition, but they also form a ring, as the set satisfies the above axioms:

$$\begin{aligned}
2\times(3\times 4)&=(2\times 3)\times 4\\
2\times(3+4)&=(2\times 3)+(2\times 4)\\
(2+3)\times 4&=(2\times 4)+(3\times 4).
\end{aligned}$$

2.8.4 Fields

Although rings support addition and multiplication, they do not necessarily support division. However, as division is such an important arithmetic operation, the *field* was created to support it, with one proviso: division by zero is not permitted. Thus we have fields of real numbers $\mathbb{R}$, rational numbers $\mathbb{Q}$, and as we shall see, the complex numbers $\mathbb{C}$. However, we will discover that quaternions do not form a field, but they do form what is called a *division ring*.

It follows that every field is a ring, but not every ring is a field.

2.8.5 Division Ring

A *division ring* or *division algebra*, is a ring in which every element has an inverse element, with the proviso that the element is non-zero. The algebra also supports non-commutative multiplication. Here is a formal description of the division ring $(S, +, \times)$:

$$
\begin{aligned}
\text{Additive associativity:}\quad & (a+b)+c = a+(b+c), & a,b,c \in S.\\
\text{Additive commutativity:}\quad & a+b = b+a, & a,b \in S.\\
\text{Additive identity } 0:\quad & 0+a = a+0, & a,0 \in S.\\
\text{Additive inverse:}\quad & a+(-a) = (-a)+a = 0, & a,-a \in S.\\
\text{Multiplicative associativity:}\quad & (a\times b)\times c = a\times(b\times c), & a,b,c \in S.\\
\text{Multiplicative identity } 1:\quad & 1\times a = a\times 1, & a,1 \in S.\\
\text{Multiplicative inverse:}\quad & a\times a^{-1} = a^{-1}\times a = 1, & a,a^{-1} \in S, a \neq 0.\\
\text{Distributivity:}\quad & a\times(b+c) = (a\times b)+(a\times c), & \text{and}\\
& (b+c)\times a = (b\times a)+(c\times a), & a,b,c \in S.
\end{aligned}
$$

In 1877 the German mathematician Ferdinand Georg Frobenius (1849–1917), proved that there are only three associative division algebras: real numbers $\mathbb{R}$, complex numbers $\mathbb{C}$, and quaternions $\mathbb{H}$.

2.9 Summary

The objective of this chapter was to remind you of the axiomatic systems underlying algebra, and how the results of arithmetic operations can be open, closed, or undefined. Perhaps some of the ideas of ordered pairs, sets, groups, fields and rings are new, and they have been included as this notation is often used in association with quaternions.

All of these ideas emerge again when we consider the algebra of complex numbers and later on, quaternions.

2.9.1 Summary of Definitions

Ordered pair
An object with two distinguishable components: $(a,\ b)$ such that $(a,\ b) \neq (b,\ a)$ unless $a = b$.

Set
Definition: A set is a collection of objects.
Notation: $k \in \mathbb{Z}$ means k belongs to the set $\mathbb{Z}$.

$\mathbb{C}$: Set of complex numbers

$\mathbb{H}$: Set of quaternions

$\mathbb{N}$: Set of natural numbers

$\mathbb{Q}$: Set of rational numbers

$\mathbb{R}$: Set of real numbers

$\mathbb{Z}$: Set of integers.

Group

Definition: A group $(S, \circ)$ is a set S and a binary operation '$\circ$' and the axioms defining closure, associativity, an identity element, and an inverse element.

$$\begin{aligned} \text{Closure: } & a \circ b \in S, & a, b \in S. \\ \text{Associativity: } & (a \circ b) \circ c = a \circ (b \circ c), & a, b, c \in S. \\ \text{Identity: } & a \circ e = e \circ a = a, & a, e \in S. \\ \text{Inverse: } & a \circ a^{-1} = a^{-1} \circ a = e, & a, a^{-1}, e \in S. \end{aligned}$$

Ring

Definition: A ring is a group whose elements can be added/subtracted *and* multiplied, using some precise axioms:

$$\begin{aligned} \text{Additive associativity: } & a + (b + c) = (a + b) + c, & a, b, c \in S. \\ \text{Multiplicative associativity: } & a \times (b \times c) = (a \times b) \times c, & a, b, c \in S. \\ \text{Distributivity: } & a \times (b + c) = (a \times b) + (a \times c), & \text{and} \\ & (a + b) \times c = (a \times c) + (b \times c), & a, b, c \in S. \end{aligned}$$

Field

Definition: A field is a ring that supports division.

Division ring

Every element of a division ring has an inverse element, with the proviso that the element is non-zero. The algebra also supports non-commutative multiplication.

$$\begin{aligned} \text{Additive associativity: } & (a + b) + c = a + (b + c), & a, b, c \in S. \\ \text{Additive commutativity: } & a + b = b + a, & a, b \in S. \\ \text{Additive identity } 0: & 0 + a = a + 0, & a, 0 \in S. \\ \text{Additive inverse: } & a + (-a) = (-a) + a = 0, & a, -a \in S. \\ \text{Multiplicative associativity: } & (a \times b) \times c = a \times (b \times c), & a, b, c \in S. \\ \text{Multiplicative identity } 1: & 1 \times a = a \times 1, & a, 1 \in S. \\ \text{Multiplicative inverse: } & a \times a^{-1} = a^{-1} \times a = 1, & a, a^{-1} \in S, a \neq 0. \\ \text{Distributivity: } & a \times (b + c) = (a \times b) + (a \times c), & \text{and} \\ & (b + c) \times a = (b \times a) + (c \times a), & a, b, c \in S. \end{aligned}$$

Reference

1. Robinson, E.: Greek and English Lexicon of the New Testament (1825). http://books.google.co.uk

Chapter 3
Complex Numbers

3.1 Introduction

In this chapter we discover how equations that have no real roots give rise to imaginary i which squares to -1. This, in turn, leads us to *complex numbers* and how they are manipulated algebraically. Many of the qualities associated with quaternions are found in complex numbers, which is why they are worthy of close examination. Readers interested in this subject may want to examine the author's book *Imaginary Mathematics for Computer Science* [1].

3.2 Imaginary Numbers

Imaginary numbers were invented to resolve problems where an equation has no real roots, such as $x^2 + 16 = 0$. The simple idea of declaring the existence of a quantity i, such that $i^2 = -1$, permits us to express the solution to this equation as

$$x = \pm 4i.$$

It is pointless trying to discover what i *really* is, i *really* is just a mathematical object that squares to -1. Nevertheless, it is a wonderful invention and does lend itself to a graphical interpretation, which we investigate in the next chapter.

In 1637 the French mathematician René Descartes (1596–1650), published *La Géométrie* [2], in which he stated that numbers incorporating $\sqrt{-1}$ were '*imaginary*', and for centuries this label has stuck. Unfortunately, it was a derogatory remark, and there is nothing *imaginary* about i—it *really* is just something that squares to -1, and when embedded within algebra creates some amazing patterns.

There doesn't appear to be any consensus concerning what to call the set of imaginary numbers—in fact it is even argued that one is unnecessary. However, if

J. Vince, *Quaternions for Computer Graphics*,
https://doi.org/10.1007/978-1-4471-7509-4_3

you decide on using one, the possibilities are $\mathbb{I}$ or $i\mathbb{R}$. Consequently, an imaginary number can be defined as

$$bi \in i\mathbb{R}, \quad i^2 = -1.$$

3.3 Powers of *i*

As $i^2 = -1$ then it should be possible to raise i to other powers. For example,

$$i^4 = i^2 i^2 = (-1)(-1) = 1$$

and

$$i^5 = ii^4 = i.$$

Therefore, we have the sequence:

i^0	i^1	i^2	i^3	i^4	i^5	i^6
1	i	-1	$-i$	1	i	-1

The cyclic pattern $(1, i, -1, -i, 1, \ldots)$ is quite striking, and reminds us of a similar pattern $(x, y, -x, -y, x, \ldots)$ which is generated by rotating around the Cartesian axes in a counter-clockwise direction. Such a similarity cannot be ignored, for when the real number line is combined with a vertical imaginary axis it gives rise to what is called the *complex plane*. But more of this in the following chapter.

The above sequence is summarised as:

$$\begin{aligned} i^{4n} &= 1 \\ i^{4n+1} &= i \\ i^{4n+2} &= -1 \\ i^{4n+3} &= -i \end{aligned}$$

where $n \in \mathbb{N}$.

But what about negative powers? Well, they, too, are also possible. Consider i^{-1}, which is evaluated as follows:

$$i^{-1} = \frac{1}{i} = \frac{1(-i)}{i(-i)} = \frac{-i}{1} = -i.$$

Similarly,

$$i^{-2} = \frac{1}{i^2} = \frac{1}{-1} = -1$$

and

$$i^{-3} = i^{-1}i^{-2} = -i(-1) = i.$$

The sequence associated with increasing negative powers is:

i^0	i^{-1}	i^{-2}	i^{-3}	i^{-4}	i^{-5}	i^{-6}
1	$-i$	-1	i	1	$-i$	-1

This time the cyclic pattern is reversed to $(1,\ -i,\ -1,\ i,\ 1,\ \ldots)$ and is similar to the pattern $(x,\ -y,\ -x,\ y,\ x,\ \ldots)$ which is generated by rotating around the Cartesian axes in a clockwise direction.

Perhaps the strangest power of all is itself: i^i, which happens to equal $e^{-\pi/2} = 0.207879576\ldots$, and is explained in Chap. 4. Having reviewed certain features of imaginary i, let's discover what happens when it's combined with real numbers.

3.4 Definition of a Complex Number

By definition, a *complex number* is the combination of a real number and an imaginary number, and is expressed as

$$z = a + bi, \quad a, b \in \mathbb{R}, \quad i^2 = -1.$$

The set of complex numbers is $\mathbb{C}$, which permits us to write $z \in \mathbb{C}$. For example, $3 + 4i$ is a complex number where 3 is the real part and $4i$ is the imaginary part. The following are all complex numbers:

$$3, \quad 3 + 4i, \quad -4 - 6i, \quad 7i, \quad 5.5 + 6.7i.$$

A real number is also a complex number—it just has no imaginary part. This leads to the idea that the set of real numbers is a subset of complex numbers, which is expressed as:

$$\mathbb{R} \subset \mathbb{C}$$

where $\subset$ means *is a subset of.*

Although some mathematicians place i before its multiplier: $i4$, others place it after the multiplier: $4i$, which is the convention used in this book. However, when i is associated with trigonometric functions, it is good practice to place it before the function to avoid any confusion with the function's angle. For example, $\sin \alpha i$ could imply that the angle is imaginary, whereas $i \sin \alpha$ implies that the value of $\sin \alpha$ is imaginary.

Therefore, a complex number can be constructed in all sorts of ways:

$$\sin\alpha + i\cos\beta, \quad 2 - i\tan\alpha, \quad 23 + x^2 i.$$

In general, we write a complex number as $a + bi$ and subject it to the normal rules of real algebra. All that we have to remember is that whenever we encounter i^2 it is replaced by -1. For example:

$$\begin{aligned}(2+3i)(3+4i) &= 2\times 3 + 2\times 4i + 3i\times 3 + 3i\times 4i\\ &= 6 + 8i + 9i + 12i^2\\ &= 6 + 17i - 12\\ &= -6 + 17i.\end{aligned}$$

3.4.1 *Addition and Subtraction of Complex Numbers*

Given two complex numbers:

$$\begin{aligned}z_1 &= a_1 + b_1 i\\ z_2 &= a_2 + b_2 i\end{aligned}$$

then,

$$z_1 \pm z_2 = (a_1 \pm a_2) + (b_1 \pm b_2)i$$

where the real and imaginary parts are added or subtracted, respectively. The operations are closed, so long as $a_1, b_1, a_2, b_2 \in \mathbb{R}$.

For example:

$$\begin{aligned}z_1 &= 2 + 3i\\ z_2 &= 4 + 2i\\ z_1 + z_2 &= 6 + 5i\\ z_1 - z_2 &= -2 + i.\end{aligned}$$

3.4.2 *Multiplying a Complex Number by a Scalar*

A complex number is multiplied by a scalar using normal algebraic rules. For example, the complex number $a + bi$ is multiplied by the scalar λ as follows:

$$\lambda(a + bi) = \lambda a + \lambda bi$$

an example is

$$3(2+5i)=6+15i.$$

3.4.3 Product of Complex Numbers

Given two complex numbers:

$$\begin{aligned} z_1 &= a_1 + b_1 i \\ z_2 &= a_2 + b_2 i \end{aligned}$$

their product is

$$\begin{aligned} z_1 z_2 &= (a_1 + b_1 i)(a_2 + b_2 i) \\ &= a_1 a_2 + a_1 b_2 i + b_1 a_2 i + b_1 b_2 i^2 \\ &= (a_1 a_2 - b_1 b_2) + (a_1 b_2 + b_1 a_2) i \end{aligned}$$

which is another complex number and confirms that the operation is closed. For example:

$$\begin{aligned} z_1 &= 3 + 4i \\ z_2 &= 3 - 2i \\ z_1 z_2 &= (3 + 4i)(3 - 2i) \\ &= 9 - 6i + 12i - 8i^2 \\ &= 9 + 6i + 8 \\ &= 17 + 6i. \end{aligned}$$

Note that the addition, subtraction and multiplication of complex numbers obey the normal axioms of algebra.

3.4.4 Square of a Complex Number

Given a complex number z, its square z^2 is given by:

$$\begin{aligned} z &= a + bi \\ z^2 &= (a + bi)(a + bi) \\ &= \left(a^2 - b^2\right) + 2abi. \end{aligned}$$

For example:

$$\begin{aligned} z &= 4 + 3i \\ z^2 &= (4 + 3i)(4 + 3i) \\ &= \left(4^2 - 3^2\right) + 2 \times 4 \times 3i \\ &= 7 + 24i. \end{aligned}$$

3.4.5 Norm of a Complex Number

The *norm*, *modulus* or *absolute value* of a complex number z is written $|z|$ and by definition is

$$\begin{aligned} z &= a + bi \\ |z| &= \sqrt{a^2 + b^2}. \end{aligned}$$

For example, the norm of $3 + 4i$ is 5. We'll see why this is so when we cover the polar representation of a complex number.

3.4.6 Complex Conjugate of a Complex Number

The product of two complex numbers, where the only difference between them is the sign of the imaginary part, gives rise to a special result:

$$\begin{aligned} (a + bi)(a - bi) &= a^2 - abi + abi - b^2 i^2 \\ &= a^2 + b^2. \end{aligned}$$

This type of product *always* results in a real quantity and is used to resolve the quotient of two complex numbers. Because this real value is such an interesting result, $a - bi$ is called the *complex conjugate* of $z = a + bi$, and is written either with a bar $\bar{z}$, or an asterisk z^*, and implies that

$$zz^* = a^2 + b^2 = |z|^2.$$

For example:

$$\begin{aligned} z &= 3 + 4i \\ z^* &= 3 - 4i \\ zz^* &= 9 + 16 = 25. \end{aligned}$$

3.4.7 Quotient of Complex Numbers

The complex conjugate provides us with a mechanism to divide one complex number by another. For instance, the quotient

$$\frac{a_1 + b_1 i}{a_2 + b_2 i}$$

is resolved by multiplying the numerator and denominator by the denominator's complex conjugate $a_2 - b_2 i$ to create a real denominator:

$$\begin{aligned}
\frac{a_1 + b_1 i}{a_2 + b_2 i} &= \frac{(a_1 + b_1 i)(a_2 - b_2 i)}{(a_2 + b_2 i)(a_2 - b_2 i)} \\
&= \frac{a_1 a_2 - a_1 b_2 i + b_1 a_2 i - b_1 b_2 i^2}{a_2^2 + b_2^2} \\
&= \left(\frac{a_1 a_2 + b_1 b_2}{a_2^2 + b_2^2}\right) + \left(\frac{b_1 a_2 - a_1 b_2}{a_2^2 + b_2^2}\right) i.
\end{aligned}$$

For example, to evaluate

$$\frac{4 + 3i}{3 + 4i}.$$

we multiply top and bottom by the complex conjugate $3 - 4i$:

$$\begin{aligned}
\frac{4 + 3i}{3 + 4i} &= \frac{(4 + 3i)(3 - 4i)}{(3 + 4i)(3 - 4i)} \\
&= \frac{12 - 16i + 9i - 12i^2}{25} \\
&= \tfrac{24}{25} - \tfrac{7}{25} i.
\end{aligned}$$

3.4.8 Inverse of a Complex Number

To compute the inverse of $z = a + bi$ we start with

$$z^{-1} = \frac{1}{z}.$$

Multiplying top and bottom by z^* we have

$$z^{-1} = \frac{z^*}{z z^*}.$$

But we have previously shown that $zz^* = |z|^2$, therefore,

$$\begin{aligned} z^{-1} &= \frac{z^*}{|z|^2} \\ &= \left(\frac{a}{a^2+b^2}\right) - \left(\frac{b}{a^2+b^2}\right)i. \end{aligned}$$

As an example, the inverse of $3+4i$ is

$$(3+4i)^{-1} = \tfrac{3}{25} - \tfrac{4}{25}i.$$

Let's test this result by multiplying $3+4i$ by its inverse:

$$(3+4i)\left(\tfrac{3}{25} - \tfrac{4}{25}i\right) = \tfrac{9}{25} - \tfrac{12}{25}i + \tfrac{12}{25}i + \tfrac{16}{25} = 1$$

which confirms the correctness of the result.

3.4.9 Square-Root of $\pm i$

To find $\sqrt{i}$ we assume that the roots are complex. Therefore, we start with

$$\begin{aligned} \sqrt{i} &= a+bi \\ i &= (a+bi)(a+bi) \\ &= a^2 + 2abi - b^2 \\ &= a^2 - b^2 + 2abi \end{aligned}$$

and equating real and imaginary parts we have

$$\begin{aligned} a^2 - b^2 &= 0 \\ 2ab &= 1. \end{aligned}$$

From this we deduce that

$$a = b = \pm\tfrac{\sqrt{2}}{2}.$$

Therefore, the roots are

$$\sqrt{i} = \pm\tfrac{\sqrt{2}}{2}(1+i).$$

Let's test this result by squaring each root to ensure the answer is i:

$$\left(\pm\tfrac{\sqrt{2}}{2}\right)^2 (1+i)(1+i) = \tfrac{1}{2}2i = i.$$

To find $\sqrt{-i}$ we assume that the roots are complex. Therefore, we start with

$$\begin{aligned}\sqrt{-i} &= a + bi\\ -i &= (a+bi)(a+bi)\\ &= a^2 + 2abi - b^2\\ &= a^2 - b^2 + 2abi\end{aligned}$$

and equating real and imaginary parts we have

$$\begin{aligned}a^2 - b^2 &= 0\\ 2ab &= -1.\end{aligned}$$

From this we deduce that

$$a = b = \pm\tfrac{\sqrt{2}}{2}i.$$

Therefore, the roots are

$$\begin{aligned}\sqrt{-i} &= \pm\tfrac{\sqrt{2}}{2}i(1+i)\\ &= \pm\tfrac{\sqrt{2}}{2}(-1+i)\\ &= \pm\tfrac{\sqrt{2}}{2}(1-i).\end{aligned}$$

Let's test this result by squaring each root to ensure the answer is $-i$:

$$\left(\pm\tfrac{\sqrt{2}}{2}\right)^2(1-i)(1-i) = -\tfrac{1}{2}2i = -i.$$

We use these roots in the next chapter to investigate the rotational properties of complex numbers.

3.5 Field Structure of Complex Numbers

The set of complex numbers $\mathbb{C}$ is a field, because it satisfies the previously defined rules for a field.

3.6 Ordered Pairs

So far, we have chosen to express a complex number as $a + bi$ where we can distinguish between the real and imaginary parts. However, one thing we cannot assume is that the real part is always first, and the imaginary part second, because $bi + a$

is also a complex number. Consequently, two functions are employed to extract the real and imaginary coefficients as follows:

$$\text{Re}(a + bi) = a$$
$$\text{Im}(a + bi) = b$$

and leads us to the idea of representing a complex number by an ordered pair where order is guaranteed:

$$a + bi = (a,\ b)$$

where b follows a to define the order. Thus the set $\mathbb{C}$ of complex numbers is equivalent to the set $\mathbb{R}^2$ of ordered pairs $(a,\ b)$.

Writing a complex number as an ordered pair was a great contribution, and first made by Hamilton in 1833. Such notation is very succinct and free from any imaginary term, which can be added whenever required.

3.6.1 Addition and Subtraction of Ordered Pairs

Given two complex numbers:

$$z_1 = a_1 + b_1 i$$
$$z_2 = a_2 + b_2 i$$

they are written as ordered pairs:

$$z_1 = (a_1,\ b_1)$$
$$z_2 = (a_2,\ b_2)$$

and

$$z_1 \pm z_2 = (a_1 \pm a_2,\ b_1 \pm b_2)$$

where the two parts are added or subtracted, respectively.

For example:

$$z_1 = 2 + 3i = (2,\ 3)$$
$$z_2 = 4 + 2i = (4,\ 2)$$
$$z_1 + z_2 = (6,\ 5)$$
$$z_1 - z_2 = (-2,\ 1).$$

3.6.2 *Multiplying an Ordered Pair by a Scalar*

We have already seen how a complex number is multiplied by a scalar, which must be the same as ordered pairs:

$$\lambda(a,\ b) = (\lambda a,\ \lambda b).$$

An example is

$$3(2,\ 5) = (6,\ 15).$$

3.6.3 *Product of Ordered Pairs*

Given two complex numbers:

$$\begin{aligned} z_1 &= a_1 + b_1 i \\ z_2 &= a_2 + b_2 i \end{aligned}$$

their product is

$$z_1 z_2 = (a_1 a_2 - b_1 b_2) + (a_1 b_2 + b_1 a_2)i$$

which must also work with ordered pairs:

$$\begin{aligned} z_1 &= (a_1,\ b_1) \\ z_2 &= (a_2,\ b_2) \\ z_1 z_2 &= (a_1,\ b_1)(a_2,\ b_2) \\ &= (a_1 a_2 - b_1 b_2,\ a_1 b_2 + b_1 a_2). \end{aligned}$$

For example:

$$\begin{aligned} z_1 &= (6,\ 2) \\ z_2 &= (4,\ 3) \\ z_1 z_2 &= (6,\ 2)(4,\ 3) \\ &= (24 - 6,\ 18 + 8) \\ &= (18,\ 26). \end{aligned}$$

3.6.4 Square of an Ordered Pair

The square of a complex number is given by:

$$\begin{aligned} z &= a + bi \\ z^2 &= (a + bi)(a + bi) \\ &= \left(a^2 - b^2\right) + 2abi. \end{aligned}$$

Therefore, the square of an ordered pair is:

$$\begin{aligned} z &= (a,\ b) \\ z^2 &= (a,\ b)(a,\ b) \\ &= \left(a^2 - b^2,\ 2ab\right). \end{aligned}$$

For example:

$$\begin{aligned} z &= (4,\ 3) \\ z^2 &= (4,\ 3)(4,\ 3) \\ &= \left(4^2 - 3^2,\ 2 \times 4 \times 3\right) \\ &= (7,\ 24). \end{aligned}$$

Let's continue to develop an algebra based upon ordered pairs that is identical to the algebra of complex numbers. We start by writing

$$\begin{aligned} z &= (a,\ b) \\ &= (a,\ 0) + (0,\ b) \\ &= a(1,\ 0) + b(0,\ 1) \end{aligned}$$

which creates the unit ordered pairs $(1,\ 0)$ and $(0,\ 1)$.

Now let's compute the product $(1,\ 0)(1,\ 0)$:

$$\begin{aligned} (1,\ 0)(1,\ 0) &= (1 - 0,\ 0) \\ &= (1,\ 0) \end{aligned}$$

which shows that $(1,\ 0)$ behaves like the real number 1. i.e. $(1,\ 0) = 1$.

Next, let's compute the product $(0,\ 1)(0,\ 1)$:

$$\begin{aligned} (0,\ 1)(0,\ 1) &= (0 - 1,\ 0) \\ &= (-1,\ 0) \end{aligned}$$

which is the real number -1:

$$(0,\ 1)^2 = -1$$

or

$$(0,\ 1) = \sqrt{-1} \quad \text{and is imaginary.}$$

This means that the ordered pair $(a,\ b)$, together with its associated rules, represents a complex number. i.e. $(a,\ b) \equiv a + bi$.

3.6.5 Norm of an Ordered Pair

The *norm*, *modulus* or *absolute value* of an ordered pair z is written $|z|$ and by definition is

$$\begin{aligned} z &= (a,\ b) \\ |z| &= \sqrt{a^2 + b^2}. \end{aligned}$$

For example, the norm of $(3,\ 4)$ is 5.

3.6.6 Complex Conjugate of an Ordered Pair

The complex conjugate of $z = a + bi$ is defined as $z^* = a - bi$, which in terms of an ordered pair is $z^* = (a,\ -b)$:

$$\begin{aligned} z &= (a,\ b) \\ z^* &= (a,\ -b) \\ zz^* &= (a,\ b)(a,\ -b) \\ &= (a^2 + b^2,\ ba - ab) \\ &= (a^2 + b^2,\ 0) \\ &= a^2 + b^2 = |z|^2. \end{aligned}$$

3.6.7 Quotient of an Ordered Pair

The technique for resolving z_1/z_2 is to multiply the expression by z_2^*/z_2^*, which using ordered pairs is

$$\begin{aligned}\frac{z_1}{z_2} &= \frac{(a_1,\ b_1)}{(a_2,\ b_2)}\\ &= \frac{(a_1,\ b_1)}{(a_2,\ b_2)}\frac{(a_2,\ -b_2)}{(a_2,\ -b_2)}\\ &= \frac{(a_1a_2 + b_1b_2,\ b_1a_2 - a_1b_2)}{(a_2^2 + b_2^2,\ 0)}\\ &= \left(\frac{a_1a_2 + b_1b_2}{a_2^2 + b_2^2},\ \frac{b_1a_2 - a_1b_2}{a_2^2 + b_2^2}\right).\end{aligned}$$

For example, to evaluate

$$\frac{(4,\ 3)}{(3,\ 4)}.$$

we multiply top and bottom by the complex conjugate $(3,\ -4)$:

$$\begin{aligned}\frac{(4,\ 3)}{(3,\ 4)} &= \frac{(4,\ 3)(3,\ -4)}{(3,\ 4)(3,\ -4)}\\ &= \left(\frac{12+12}{25},\ \frac{9-16}{25}\right)\\ &= \left(\tfrac{24}{25},\ -\tfrac{7}{25}\right).\end{aligned}$$

3.6.8 *Inverse of an Ordered Pair*

We have previously shown that z^{-1} is

$$z^{-1} = \frac{z^*}{zz^*} = \frac{z^*}{|z|^2}$$

which using ordered pairs is

$$\begin{aligned}z &= (a,\ b)\\ z^{-1} &= \frac{(a,\ -b)}{(a,\ b)(a,\ -b)}\\ &= \frac{(a,\ -b)}{(a^2 + b^2,\ 0)}\\ &= \left(\frac{a}{a^2 + b^2},\ \frac{-b}{a^2 + b^2}\right).\end{aligned}$$

As an illustration, the inverse of $(3,\ 4)$ is

$$(3,\ 4)^{-1} = \left(\tfrac{3}{25},\ -\tfrac{4}{25}\right).$$

Let's test this result by multiplying (3, 4) by its inverse:

$$\begin{aligned}(3,\ 4)\left(\tfrac{3}{25},\ -\tfrac{4}{25}\right) &= \left(\tfrac{9}{25}+\tfrac{16}{25},\ -\tfrac{12}{25}+\tfrac{12}{25}\right)\\ &= (1,\ 0).\end{aligned}$$

3.6.9 *Square-Root of* $\pm i$

To find $\sqrt{i}$ we assume that the roots are complex. Therefore, we start with

$$\begin{aligned}\sqrt{i} &= (a,\ b)\\ i &= (a,\ b)(a,\ b)\\ (0,\ 1) &= \left(a^2-b^2,\ 2ab\right)\end{aligned}$$

and equating left and right ordered elements we have

$$\begin{aligned}a^2-b^2 &= 0\\ 2ab &= 1.\end{aligned}$$

From this we deduce that

$$a=b=\pm\tfrac{\sqrt{2}}{2}.$$

Therefore, the roots are

$$\sqrt{i}=\pm\tfrac{\sqrt{2}}{2}(1,\ 1).$$

Let's test this result by squaring each root to ensure the answer is i:

$$\left(\pm\tfrac{\sqrt{2}}{2}\right)^2(1,\ 1)(1,\ 1)=\tfrac{1}{2}(0,\ 2)=(0,\ 1).$$

To find $\sqrt{-i}$ we assume that the roots are complex. Therefore, we start with

$$\begin{aligned}\sqrt{-i} &= (a,\ b)\\ -i &= (a,\ b)(a,\ b)\\ (0,\ -1) &= \left(a^2-b^2,\ 2ab\right)\end{aligned}$$

and equating left and right ordered elements we have

$$\begin{aligned}a^2-b^2 &= 0\\ 2ab &= -1.\end{aligned}$$

From this we deduce that

$$\begin{aligned} a = b &= \pm\tfrac{\sqrt{2}}{2} i \\ &= \pm\tfrac{\sqrt{2}}{2}(0,\ 1)(1,\ 1) \\ &= \pm\tfrac{\sqrt{2}}{2}(-1,\ 1). \end{aligned}$$

Therefore, the roots are

$$\sqrt{-i} = \pm\tfrac{\sqrt{2}}{2}(1,\ -1).$$

Let's test this result by squaring each root to ensure the answer is $-i$:

$$\left(\pm\tfrac{\sqrt{2}}{2}\right)^2 (1,\ -1)(1,\ -1) = \tfrac{1}{2}(0,\ -2) = (0,\ -1).$$

It is obvious from the above definitions that ordered pairs provide an alternative notation for expressing complex numbers, where the imaginary feature is embedded within the product axiom. We will also use ordered pairs to define a quaternion with three imaginary terms, which when incorporated within the product axiom remain hidden.

3.7 Matrix Representation of a Complex Number

As quaternions have a matrix representation, perhaps we should investigate the matrix representation for a complex number.

Although I have only hinted that i can be regarded as some sort of rotational operator, this is the perfect way of visualising it. In Chap. 4 we discover that multiplying a complex number by i effectively rotates the number 90° anti-clockwise. So for the moment, it can be represented by a rotation matrix of 90°:

$$i \equiv \begin{bmatrix} \cos 90^\circ & -\sin 90^\circ \\ \sin 90^\circ & \cos 90^\circ \end{bmatrix} = \begin{bmatrix} 0 & -1 \\ 1 & 0 \end{bmatrix}$$

and the 2×2 identity matrix is

$$\begin{bmatrix} 1 & 0 \\ 0 & 1 \end{bmatrix}.$$

This permits us to write a complex number as:

$$\begin{aligned}a+bi &= a\begin{bmatrix}1 & 0\\0 & 1\end{bmatrix}+b\begin{bmatrix}0 & -1\\1 & 0\end{bmatrix}\\&=\begin{bmatrix}a & 0\\0 & a\end{bmatrix}+\begin{bmatrix}0 & -b\\b & 0\end{bmatrix}\\&=\begin{bmatrix}a & -b\\b & a\end{bmatrix}.\end{aligned}$$

Note that the matrix representing i squares to -1:

$$\begin{aligned}\begin{bmatrix}0 & -1\\1 & 0\end{bmatrix}\begin{bmatrix}0 & -1\\1 & 0\end{bmatrix}&=\begin{bmatrix}-1 & 0\\0 & -1\end{bmatrix}\\&=-1\begin{bmatrix}1 & 0\\0 & 1\end{bmatrix}.\end{aligned}$$

However, we must also remember that $i^2=(-i)^2=-1$, which is interpreted as anti-clockwise and clockwise rotations in the complex plane. This is confirmed by:

$$\begin{aligned}\begin{bmatrix}0 & 1\\-1 & 0\end{bmatrix}\begin{bmatrix}0 & 1\\-1 & 0\end{bmatrix}&=\begin{bmatrix}-1 & 0\\0 & -1\end{bmatrix}\\&=-1\begin{bmatrix}1 & 0\\0 & 1\end{bmatrix}.\end{aligned}$$

Now let's employ matrix notation for all the arithmetic operations used for complex numbers.

3.7.1 Adding and Subtracting Complex Numbers

Two complex numbers are added or subtracted as follows:

$$\begin{aligned}z_1 &= a_1+b_1i\\z_2 &= a_2+b_2i\\z_1 &= \begin{bmatrix}a_1 & -b_1\\b_1 & a_1\end{bmatrix}\\z_2 &= \begin{bmatrix}a_2 & -b_2\\b_2 & a_2\end{bmatrix}\\z_1\pm z_2 &= \begin{bmatrix}a_1 & -b_1\\b_1 & a_1\end{bmatrix}\pm\begin{bmatrix}a_2 & -b_2\\b_2 & a_2\end{bmatrix}\\&=\begin{bmatrix}a_1\pm a_2 & -(b_1\pm b_2)\\b_1\pm b_2 & a_1\pm a_2\end{bmatrix}.\end{aligned}$$

For example:

$$
\begin{aligned}
z_1 &= 2 + 3i \\
z_2 &= 4 + 2i \\
z_1 &= \begin{bmatrix} 2 & -3 \\ 3 & 2 \end{bmatrix} \\
z_2 &= \begin{bmatrix} 4 & -2 \\ 2 & 4 \end{bmatrix} \\
z_1 \pm z_2 &= \begin{bmatrix} 2 & -3 \\ 3 & 2 \end{bmatrix} \pm \begin{bmatrix} 4 & -2 \\ 2 & 4 \end{bmatrix} \\
z_1 + z_2 &= \begin{bmatrix} 6 & -5 \\ 5 & 6 \end{bmatrix} = 6 + 5i \\
z_1 - z_2 &= \begin{bmatrix} -2 & -1 \\ 1 & -2 \end{bmatrix} = -2 + i.
\end{aligned}
$$

3.7.2 Product of Two Complex Numbers

The product of two complex numbers is computed as follows:

$$
\begin{aligned}
z_1 &= a_1 + b_1 i \\
z_2 &= a_2 + b_2 i \\
z_1 z_2 &= \begin{bmatrix} a_1 & -b_1 \\ b_1 & a_1 \end{bmatrix} \begin{bmatrix} a_2 & -b_2 \\ b_2 & a_2 \end{bmatrix} \\
&= \begin{bmatrix} a_1 a_2 - b_1 b_2 & -(a_1 b_2 + b_1 a_2) \\ a_1 b_2 + b_1 a_2 & a_1 a_2 - b_1 b_2 \end{bmatrix}.
\end{aligned}
$$

For example:

$$
\begin{aligned}
z_1 &= 6 + 2i \\
z_2 &= 4 + 3i \\
z_1 z_2 &= \begin{bmatrix} 6 & -2 \\ 2 & 6 \end{bmatrix} \begin{bmatrix} 4 & -3 \\ 3 & 4 \end{bmatrix} \\
&= \begin{bmatrix} 24 - 6 & -(18 + 8) \\ 18 + 8 & 24 - 6 \end{bmatrix} \\
&= \begin{bmatrix} 18 & -26 \\ 26 & 18 \end{bmatrix}.
\end{aligned}
$$

3.7.3 Norm Squared of a Complex Number

The square of the norm is as the determinant of the matrix:

$$
\begin{aligned}
z &= a + bi \\
&= \begin{bmatrix} a & -b \\ b & a \end{bmatrix} \\
|z|^2 = a^2 + b^2 &= \begin{vmatrix} a & -b \\ b & a \end{vmatrix}.
\end{aligned}
$$

3.7.4 Complex Conjugate of a Complex Number

The complex conjugate of a complex number is

$$
\begin{aligned}
z = a + bi &= \begin{bmatrix} a & -b \\ b & a \end{bmatrix} \\
z^* = a - bi &= \begin{bmatrix} a & b \\ -b & a \end{bmatrix}.
\end{aligned}
$$

The product $zz^* = a^2 + b^2$:

$$
\begin{aligned}
zz^* &= \begin{bmatrix} a & -b \\ b & a \end{bmatrix} \begin{bmatrix} a & b \\ -b & a \end{bmatrix} \\
&= \begin{bmatrix} a^2 + b^2 & 0 \\ 0 & a^2 + b^2 \end{bmatrix} \\
&= \left(a^2 + b^2\right) \begin{bmatrix} 1 & 0 \\ 0 & 1 \end{bmatrix}.
\end{aligned}
$$

For example:

$$
\begin{aligned}
z = 3 + 4i &= \begin{bmatrix} 3 & -4 \\ 4 & 3 \end{bmatrix} \\
z^* = 3 - 4i &= \begin{bmatrix} 3 & 4 \\ -4 & 3 \end{bmatrix} \\
zz^* &= \begin{bmatrix} 3 & -4 \\ 4 & 3 \end{bmatrix} \begin{bmatrix} 3 & 4 \\ -4 & 3 \end{bmatrix} = \begin{bmatrix} 25 & 0 \\ 0 & 25 \end{bmatrix} \\
&= 25 \begin{bmatrix} 1 & 0 \\ 0 & 1 \end{bmatrix}.
\end{aligned}
$$

3.7.5 Inverse of a Complex Number

The inverse of 2 × 2 matrix **A** is given by

$$\mathbf{A} = \begin{bmatrix} a_{11} & a_{12} \\ a_{21} & a_{22} \end{bmatrix}$$

$$\mathbf{A}^{-1} = \frac{1}{a_{11}a_{22} - a_{12}a_{21}} \begin{bmatrix} a_{22} & -a_{12} \\ -a_{21} & a_{12} \end{bmatrix}$$

therefore, the inverse of z is given by

$$z = a + bi$$

$$z = \begin{bmatrix} a & -b \\ b & a \end{bmatrix}$$

$$z^{-1} = \frac{1}{a^2 + b^2} \begin{bmatrix} a & b \\ -b & a \end{bmatrix}.$$

For example:

$$z = 3 + 4i$$

$$z = \begin{bmatrix} 3 & -4 \\ 4 & 3 \end{bmatrix}$$

$$z^{-1} = \frac{1}{25} \begin{bmatrix} 3 & 4 \\ -4 & 3 \end{bmatrix}.$$

3.7.6 Quotient of a Complex Number

The quotient of two complex numbers is computed as follows:

$$z_1 = a_1 + b_1 i$$

$$z_2 = a_2 + b_2 i$$

$$\frac{z_1}{z_2} = z_1 z_2^{-1}$$

$$= \begin{bmatrix} a_1 & -b_1 \\ b_1 & a_1 \end{bmatrix} \frac{1}{a_2^2 + b_2^2} \begin{bmatrix} a_2 & b_2 \\ -b_2 & a_2 \end{bmatrix}$$

$$= \frac{1}{a_2^2 + b_2^2} \begin{bmatrix} a_1 a_2 + b_1 b_2 & -(b_1 a_2 - a_1 b_2) \\ b_1 a_2 - a_1 b_2 & a_1 a_2 + b_1 b_2 \end{bmatrix}.$$

For example:

$$
\begin{aligned}
z_1 &= 4+3i\\
z_2 &= 3+4i\\
\frac{z_1}{z_2} &= z_1 z_2^{-1}\\
&= \begin{bmatrix} 4 & -3\\ 3 & 4 \end{bmatrix} \frac{1}{9+16} \begin{bmatrix} 3 & 4\\ -4 & 3 \end{bmatrix}\\
&= \tfrac{1}{25} \begin{bmatrix} 24 & 7\\ -7 & 24 \end{bmatrix}.
\end{aligned}
$$

3.7.7 Square-Root of $\pm i$

To find $\sqrt{i}$ we assume that the roots are complex. Therefore, we start with

$$
\begin{aligned}
\sqrt{i} &= \begin{bmatrix} a & -b\\ b & a \end{bmatrix}\\
i &= \begin{bmatrix} a & -b\\ b & a \end{bmatrix} \begin{bmatrix} a & -b\\ b & a \end{bmatrix}\\
\begin{bmatrix} 0 & -1\\ 1 & 0 \end{bmatrix} &= \begin{bmatrix} a^2-b^2 & -2ab\\ 2ab & a^2-b^2 \end{bmatrix}
\end{aligned}
$$

and equating left and right matrices we have

$$
\begin{aligned}
a^2 - b^2 &= 0\\
2ab &= 1.
\end{aligned}
$$

From this we deduce that

$$
a = b = \pm\tfrac{\sqrt{2}}{2}.
$$

Therefore, the roots are

$$
\sqrt{i} = \pm\tfrac{\sqrt{2}}{2} \begin{bmatrix} 1 & -1\\ 1 & 1 \end{bmatrix}.
$$

Let's test this result by squaring each root to ensure the answer is i:

$$
\left(\pm\tfrac{\sqrt{2}}{2}\right)^2 \begin{bmatrix} 1 & -1\\ 1 & 1 \end{bmatrix} \begin{bmatrix} 1 & -1\\ 1 & 1 \end{bmatrix} = \tfrac{1}{2} \begin{bmatrix} 0 & -2\\ 2 & 0 \end{bmatrix} = i
$$

To find $\sqrt{-i}$ we assume that the roots are complex. Therefore, we start with

$$
\begin{aligned}
\sqrt{-i} &= \begin{bmatrix} a & -b \\ b & a \end{bmatrix} \\
-i &= \begin{bmatrix} a & -b \\ b & a \end{bmatrix}\begin{bmatrix} a & -b \\ b & a \end{bmatrix} \\
\begin{bmatrix} 0 & 1 \\ -1 & 0 \end{bmatrix} &= \begin{bmatrix} a^2-b^2 & -2ab \\ 2ab & a^2-b^2 \end{bmatrix}
\end{aligned}
$$

and equating left and right matrices we have

$$
\begin{aligned}
a^2 - b^2 &= 0 \\
2ab &= -1.
\end{aligned}
$$

From this we deduce that

$$a = b = \pm\tfrac{\sqrt{2}}{2}i.$$

Therefore, the roots are

$$\sqrt{-i} = \pm\tfrac{\sqrt{2}}{2}\begin{bmatrix} 0 & -1 \\ 1 & 0 \end{bmatrix}\begin{bmatrix} 1 & -1 \\ 1 & 1 \end{bmatrix} = \pm\tfrac{\sqrt{2}}{2}\begin{bmatrix} 1 & 1 \\ -1 & 1 \end{bmatrix}.$$

Let's test this result by squaring each root to ensure the answer is i:

$$\left(\pm\tfrac{\sqrt{2}}{2}\right)^2\begin{bmatrix} 1 & 1 \\ -1 & 1 \end{bmatrix}\begin{bmatrix} 1 & 1 \\ -1 & 1 \end{bmatrix} = \tfrac{1}{2}\begin{bmatrix} 0 & 2 \\ -2 & 0 \end{bmatrix} = -i$$

3.8 Summary

We have shown in this chapter that the set of complex numbers is a field, as they satisfy the requirement for closure, associativity, distributivity, an identity element, and an inverse. We have also shown that there is a one-to-one correspondence between a complex number and an ordered pair, and that a complex number can be represented as a matrix, which permits us to compute all complex number operations as matrix operations or ordered pairs.

If this the first time you have come across complex numbers you probably will have found them strange on the one hand, and amazing on the other. Simply by declaring the existence of i that squares to -1, opens up a new number system that unifies large areas of mathematics.

3.8.1 Summary of Definitions

Definition

$i\mathbb{R}$ is the set of imaginary numbers: $ib \in i\mathbb{R}, \quad i^2 = -1$.

Complex Number

$$\text{Real unit: } 1$$
$$\text{Imaginary unit: } i.$$

Ordered Pair

$$1 = (1,\ 0)$$
$$i = (0,\ 1).$$

Matrix

$$1 = \begin{bmatrix} 1 & 0 \\ 0 & 1 \end{bmatrix}$$
$$i = \begin{bmatrix} 0 & -1 \\ 1 & 0 \end{bmatrix}.$$

$\mathbb{C}$ is the set of complex numbers: $z = a + ib, \quad a \in \mathbb{R}, \quad ib \in i\mathbb{R}, \quad z \in \mathbb{C}$.

Complex Number

$$z = a + bi.$$

Ordered Pair

$$z = (a,\ b).$$

Matrix

$$a + bi = \begin{bmatrix} a & -b \\ b & a \end{bmatrix}.$$

Addition and Subtraction

Complex Number

$$z_1 = a_1 + b_1 i$$
$$z_2 = a_2 + b_2 i$$
$$z_1 \pm z_2 = a_1 \pm a_2 + (b_1 \pm b_2)i.$$

Ordered Pair

$$
\begin{aligned}
z_1 &= (a_1,\ b_1)\\
z_2 &= (a_2,\ b_2)\\
z_1 \pm z_2 &= (a_1 \pm a_2,\ b_1 \pm b_2).
\end{aligned}
$$

Matrix

$$
\begin{aligned}
z_1 &= \begin{bmatrix} a_1 & -b_1 \\ b_1 & a_1 \end{bmatrix}\\
z_2 &= \begin{bmatrix} a_2 & -b_2 \\ b_2 & a_2 \end{bmatrix}\\
z_1 \pm z_2 &= \begin{bmatrix} a_1 & -b_1 \\ b_1 & a_1 \end{bmatrix} \pm \begin{bmatrix} a_2 & -b_2 \\ b_2 & a_2 \end{bmatrix}\\
&= \begin{bmatrix} a_1 \pm a_2 & -(b_1 \pm b_2) \\ b_1 \pm b_2 & a_1 \pm a_2 \end{bmatrix}.
\end{aligned}
$$

Product

Complex Number

$$
\begin{aligned}
z_1 &= a_1 + b_1 i\\
z_2 &= a_2 + b_2 i\\
z_1 z_2 &= (a_1 + b_1 i)(a_2 + b_2 i)\\
&= (a_1 a_2 - b_1 b_2) + (a_1 b_2 + b_1 a_2)i.
\end{aligned}
$$

Ordered Pair

$$
\begin{aligned}
z_1 &= (a_1,\ b_1)\\
z_2 &= (a_2,\ b_2)\\
z_1 z_2 &= (a_1,\ b_1)(a_2,\ b_2)\\
&= (a_1 a_2 - b_1 b_2,\ a_1 b_2 + b_1 a_2).
\end{aligned}
$$

Matrix

$$
\begin{aligned}
z_1 &= \begin{bmatrix} a_1 & -b_1 \\ b_1 & a_1 \end{bmatrix}\\
z_2 &= \begin{bmatrix} a_2 & -b_2 \\ b_2 & a_2 \end{bmatrix}
\end{aligned}
$$

$$\begin{aligned} z_1 z_2 &= \begin{bmatrix} a_1 & -b_1 \\ b_1 & a_1 \end{bmatrix} \begin{bmatrix} a_2 & -b_2 \\ b_2 & a_2 \end{bmatrix} \\ &= \begin{bmatrix} a_1 a_2 - b_1 b_2 & -(a_1 b_2 + b_1 a_2) \\ a_1 b_2 + b_1 a_2 & a_1 a_2 - b_1 b_2 \end{bmatrix}. \end{aligned}$$

Square

Complex Number

$$\begin{aligned} z &= a + bi \\ z^2 &= (a + bi)\,(a + bi) \\ &= \left(a^2 - b^2\right) + 2abi. \end{aligned}$$

Ordered Pair

$$\begin{aligned} z &= (a,\ b) \\ z^2 &= (a,\ b)(a,\ b) \\ &= \left(a^2 - b^2,\ 2ab\right). \end{aligned}$$

Matrix

$$\begin{aligned} z &= \begin{bmatrix} a & -b \\ b & a \end{bmatrix} \\ z^2 &= \begin{bmatrix} a & -b \\ b & a \end{bmatrix} \begin{bmatrix} a & -b \\ b & a \end{bmatrix} \\ &= \begin{bmatrix} a^2 - b^2 & -2ab \\ 2ab & a^2 - b^2 \end{bmatrix}. \end{aligned}$$

Norm

Complex Number

$$\begin{aligned} z &= a + bi \\ |z| &= \sqrt{a^2 + b^2}. \end{aligned}$$

Ordered Pair

$$\begin{aligned} z &= (a,\ b) \\ |z| &= \sqrt{a^2 + b^2}. \end{aligned}$$

Matrix

$$z = \begin{bmatrix} a & -b \\ b & a \end{bmatrix}$$
$$|z|^2 = \begin{vmatrix} a & -b \\ b & a \end{vmatrix}.$$

Complex Conjugate

Complex Number

$$z = a + bi$$
$$z^* = a - bi.$$

Ordered Pair

$$z = (a,\ b)$$
$$z^* = (a,\ -b).$$

Matrix

$$z = \begin{bmatrix} a & -b \\ b & a \end{bmatrix}$$
$$z^* = \begin{bmatrix} a & b \\ -b & a \end{bmatrix}.$$

Inverse

Complex Number

$$\begin{aligned} z &= a + bi \\ z^{-1} &= \frac{z^*}{|z|^2} \\ &= \left(\frac{a}{a^2 + b^2}\right) - \left(\frac{b}{a^2 + b^2}\right) i. \end{aligned}$$

Ordered Pair

$$\begin{aligned} z &= (a,\ b) \\ z^* &= (a,\ -b) \\ |z|^2 &= a^2 + b^2 \end{aligned}$$

$$\frac{1}{z} = z^{-1} = \frac{z^*}{|z|^2}$$
$$= \left(\frac{a}{a^2+b^2},\ \frac{-b}{a^2+b^2}\right).$$

Matrix

$$z = \begin{bmatrix} a & -b \\ b & a \end{bmatrix}$$
$$z^* = \begin{bmatrix} a & b \\ -b & a \end{bmatrix}$$
$$|z|^2 = a^2 + b^2$$
$$\frac{1}{z} = z^{-1} = \frac{z^*}{|z|^2}$$
$$= \frac{1}{a^2+b^2}\begin{bmatrix} a & b \\ -b & a \end{bmatrix}.$$

Quotient

Complex Number

$$z_1 = a_1 + b_1 i$$
$$z_2 = a_2 + b_2 i$$
$$\frac{z_1}{z_2} = \frac{a_1 + b_1 i}{a_2 + b_2 i}$$
$$= \left(\frac{a_1a_2 + b_1b_2}{a_2^2 + b_2^2}\right) + \left(\frac{b_1a_2 - a_1b_2}{a_2^2 + b_2^2}\right)i.$$

Ordered Pair

$$z_1 = (a_1,\ b_1)$$
$$z_2 = (a_2,\ b_2)$$
$$\frac{z_1}{z_2} = \frac{(a_1,\ b_1)}{(a_2,\ b_2)}$$
$$= \left(\frac{a_1a_2 + b_1b_2}{a_2^2 + b_2^2},\ \frac{b_1a_2 - a_1b_2}{a_2^2 + b_2^2}\right).$$

Matrix

$$
\begin{aligned}
z_1 &= \begin{bmatrix} a_1 & -b_1 \\ b_1 & a_1 \end{bmatrix} \\
z_2 &= \begin{bmatrix} a_2 & -b_2 \\ b_2 & a_2 \end{bmatrix} \\
\frac{z_1}{z_2} &= z_1 z_2^{-1} \\
&= \frac{1}{a_2^2 + b_2^2} \begin{bmatrix} a_1 & -b_1 \\ b_1 & a_1 \end{bmatrix} \begin{bmatrix} a_2 & b_2 \\ -b_2 & a_2 \end{bmatrix} \\
&= \frac{1}{a_2^2 + b_2^2} \begin{bmatrix} a_1a_2 + b_1b_2 & -(b_1a_2 - a_1b_2) \\ b_1a_2 - a_1b_2 & a_1a_2 + b_1b_2 \end{bmatrix}.
\end{aligned}
$$

Square root of $\pm i$

Complex Number

$$
\begin{aligned}
\sqrt{i} &= \pm\tfrac{\sqrt{2}}{2}(1 + i) \\
\sqrt{-i} &= \pm\tfrac{\sqrt{2}}{2}(1 - i).
\end{aligned}
$$

Ordered Pair

$$
\begin{aligned}
\sqrt{i} &= \pm\tfrac{\sqrt{2}}{2}(1,\ 1) \\
\sqrt{-i} &= \pm\tfrac{\sqrt{2}}{2}(1,\ -1).
\end{aligned}
$$

Matrix

$$
\begin{aligned}
\sqrt{i} &= \pm\tfrac{\sqrt{2}}{2} \begin{bmatrix} 1 & -1 \\ 1 & 1 \end{bmatrix} \\
\sqrt{-i} &= \pm\tfrac{\sqrt{2}}{2} \begin{bmatrix} 1 & 1 \\ -1 & 1 \end{bmatrix}.
\end{aligned}
$$

3.9 Worked Examples

Here are some further worked examples that employ the ideas described above. In some cases a test is included to confirm the result.

3.9.1 Adding and Subtracting Complex Numbers

Add and subtract z_1 and z_2.

Complex Number

$$\begin{aligned} z_1 &= 12 + 6i \\ z_2 &= 10 - 4i \\ z_1 + z_2 &= 22 + 2i \\ z_1 - z_2 &= 2 + 10i. \end{aligned}$$

Ordered Pair

$$\begin{aligned} z_1 &= (12,\ 6) \\ z_2 &= (10,\ -4) \\ z_1 + z_2 &= (12,\ 6) + (10,\ -4) = (22,\ 2) \\ z_1 - z_2 &= (12,\ 6) - (10,\ -4) = (2,\ 10). \end{aligned}$$

Matrix

$$\begin{aligned} z_1 &= \begin{bmatrix} 12 & -6 \\ 6 & 12 \end{bmatrix} \\ z_2 &= \begin{bmatrix} 10 & 4 \\ -4 & 10 \end{bmatrix} \\ z_1 + z_2 &= \begin{bmatrix} 12 & -6 \\ 6 & 12 \end{bmatrix} + \begin{bmatrix} 10 & 4 \\ -4 & 10 \end{bmatrix} = \begin{bmatrix} 22 & -2 \\ 2 & 22 \end{bmatrix} \\ z_1 - z_2 &= \begin{bmatrix} 12 & -6 \\ 6 & 12 \end{bmatrix} - \begin{bmatrix} 10 & 4 \\ -4 & 10 \end{bmatrix} = \begin{bmatrix} 2 & -10 \\ 10 & 2 \end{bmatrix}. \end{aligned}$$

3.9.2 Product of Complex Numbers

Compute the product $z_1 z_2$.

Complex Number

$$\begin{aligned} z_1 &= 12 + 6i \\ z_2 &= 10 - 4i \\ z_1 z_2 &= (12 + 6i)(10 - 4i) \\ &= 144 + 12i. \end{aligned}$$

Ordered Pair

$$\begin{aligned} z_1 &= (12,\ 6) \\ z_2 &= (10,\ -4) \\ z_1 z_2 &= (12,\ 6)(10,\ -4) \\ &= (120 + 24,\ -48 + 60) \\ &= (144,\ 12). \end{aligned}$$

Matrix

$$\begin{aligned} z_1 &= \begin{bmatrix} 12 & -6 \\ 6 & 12 \end{bmatrix} \\ z_2 &= \begin{bmatrix} 10 & 4 \\ -4 & 10 \end{bmatrix} \\ z_1 z_2 &= \begin{bmatrix} 12 & -6 \\ 6 & 12 \end{bmatrix} \begin{bmatrix} 10 & 4 \\ -4 & 10 \end{bmatrix} = \begin{bmatrix} 144 & -12 \\ 12 & 144 \end{bmatrix}. \end{aligned}$$

3.9.3 Multiplying a Complex Number by i

Multiply z_1 by i.

Complex Number

$$\begin{aligned} z_1 &= 12 + 6i \\ z_1 i &= (12 + 6i)i \\ &= -6 + 12i. \end{aligned}$$

Ordered Pair

$$\begin{aligned} z_1 &= (12,\ 6) \\ i &= (0,\ 1) \\ z_1 i &= (12,\ 6)(0,\ 1) \\ &= (-6,\ 12). \end{aligned}$$

Matrix

$$
\begin{aligned}
z_1 &= \begin{bmatrix} 12 & -6 \\ 6 & 12 \end{bmatrix} \\
i &= \begin{bmatrix} 0 & -1 \\ 1 & 0 \end{bmatrix} \\
z_1 z_2 &= \begin{bmatrix} 12 & -6 \\ 6 & 12 \end{bmatrix} \begin{bmatrix} 0 & -1 \\ 1 & 0 \end{bmatrix} = \begin{bmatrix} -6 & -12 \\ 12 & -6 \end{bmatrix}.
\end{aligned}
$$

3.9.4 The Norm of a Complex Number

Compute the norm of z_1.

Complex Number

$$
\begin{aligned}
z_1 &= 12 + 6i \\
|z_1| &= \sqrt{12^2 + 6^2} \approx 13.416.
\end{aligned}
$$

Ordered Pair

$$
\begin{aligned}
z_1 &= (12,\ 6) \\
|z_1| &= \sqrt{12^2 + 6^2} \approx 13.416.
\end{aligned}
$$

Matrix

$$
\begin{aligned}
z_1 &= \begin{bmatrix} 12 & -6 \\ 6 & 12 \end{bmatrix} \\
|z_1| &= \begin{vmatrix} 12 & -6 \\ 6 & 12 \end{vmatrix} = \sqrt{12^2 + 6^2} \approx 13.416.
\end{aligned}
$$

3.9.5 The Complex Conjugate of a Complex Number

Compute the complex conjugate of the following.

Complex Number

$$
\begin{aligned}
(2 + 3i)^* &= 2 - 3i \\
1^* &= 1 \\
i^* &= -i.
\end{aligned}
$$

Ordered Pair

$$(2,\ 3)^* = (2,\ -3)$$
$$(1,\ 0)^* = (1,\ 0)$$
$$(0,\ 1)^* = (0,\ -1).$$

Matrix

$$z = \begin{bmatrix} 2 & -3 \\ 3 & 2 \end{bmatrix}$$
$$z^* = \begin{bmatrix} 2 & 3 \\ -3 & 2 \end{bmatrix}$$
$$1 = \begin{bmatrix} 1 & 0 \\ 0 & 1 \end{bmatrix}$$
$$1^* = \begin{bmatrix} 1 & 0 \\ 0 & 1 \end{bmatrix}$$
$$i = \begin{bmatrix} 0 & -1 \\ 1 & 0 \end{bmatrix}$$
$$i^* = \begin{bmatrix} 0 & 1 \\ -1 & 0 \end{bmatrix}.$$

3.9.6 The Quotient of Two Complex Numbers

Compute the quotient $(2 + 3i)/(3 + 4i)$.

Complex Number

$$\begin{aligned} \frac{2+3i}{3+4i} &= \frac{(2+3i)}{(3+4i)}\frac{(3-4i)}{(3-4i)} \\ &= \frac{6-8i+9i+12}{25} \\ &= \tfrac{18}{25} + \tfrac{1}{25}i. \end{aligned}$$

Test

$$\begin{aligned} (3+4i)\left(\tfrac{18}{25} + \tfrac{1}{25}i\right) &= \tfrac{54}{25} + \tfrac{3}{25}i + \tfrac{72}{25}i - \tfrac{4}{25} \\ &= 2 + 3i. \end{aligned}$$

Ordered Pair

$$\begin{aligned}\frac{(2,\ 3)}{(3,\ 4)} &= \frac{(2,\ 3)}{(3,\ 4)}\frac{(3,\ -4)}{(3,\ -4)}\\ &= \frac{(6+12,\ 1)}{(9+16,\ 0)}\\ &= \left(\tfrac{18}{25},\ \tfrac{1}{25}\right).\end{aligned}$$

Matrix

$$\begin{aligned}z_1 &= \begin{bmatrix} 2 & -3 \\ 3 & 2 \end{bmatrix}\\ z_2 &= \begin{bmatrix} 3 & -4 \\ 4 & 3 \end{bmatrix}\\ \frac{z_1}{z_2} &= z_1 z_2^{-1}\\ &= \tfrac{1}{25}\begin{bmatrix} 2 & -3 \\ 3 & 2 \end{bmatrix}\begin{bmatrix} 3 & 4 \\ -4 & 3 \end{bmatrix}\\ &= \tfrac{1}{25}\begin{bmatrix} 18 & -1 \\ 1 & 18 \end{bmatrix}.\end{aligned}$$

3.9.7 Divide a Complex Number by i

Divide $2 + 3i$ by i.

Complex Number

$$\begin{aligned}\frac{2+3i}{0+i} &= \frac{(2+3i)}{(0+i)}\frac{(0-i)}{(0-i)}\\ &= \frac{-2i+3}{1}\\ &= 3-2i.\end{aligned}$$

Test

$$i(3-2i) = 2+3i.$$

Ordered Pair

$$\begin{aligned}\frac{(2,\ 3)}{(0,\ 1)} &= \frac{(2,\ 3)}{(0,\ 1)}\frac{(0,\ -1)}{(0,\ -1)}\\ &= \frac{(3,\ -2)}{(1,\ 0)}\\ &= (3,\ -2).\end{aligned}$$

Matrix

$$\begin{aligned}z &= \begin{bmatrix} 2 & -3 \\ 3 & 2 \end{bmatrix}\\ i &= \begin{bmatrix} 0 & -1 \\ 1 & 0 \end{bmatrix}\\ i^{-1} &= \begin{bmatrix} 0 & 1 \\ -1 & 0 \end{bmatrix}\\ zi^{-1} &= \begin{bmatrix} 2 & -3 \\ 3 & 2 \end{bmatrix}\begin{bmatrix} 0 & 1 \\ -1 & 0 \end{bmatrix} = \begin{bmatrix} 3 & 2 \\ -2 & 3 \end{bmatrix}.\end{aligned}$$

3.9.8 Divide a Complex Number by −i

Divide $2 + 3i$ by $-i$.

Complex Number

$$\begin{aligned}\frac{2+3i}{0-i} &= \frac{(2+3i)}{(0-i)}\frac{(0+i)}{(0+i)}\\ &= \frac{2i-3}{1}\\ &= -3+2i.\end{aligned}$$

Test

$$-i(-3+2i) = 2+3i.$$

Ordered Pair

$$\begin{aligned}\frac{(2,\ 3)}{(0,\ -1)} &= \frac{(2,\ 3)}{(0,\ -1)}\frac{(0,\ 1)}{(0,\ 1)}\\ &= \frac{(-3,\ 2)}{1}\\ &= (-3,\ 2).\end{aligned}$$

Matrix

$$
\begin{aligned}
z &= \begin{bmatrix} 2 & -3 \\ 3 & 2 \end{bmatrix} \\
-i &= \begin{bmatrix} 0 & 1 \\ -1 & 0 \end{bmatrix} \\
-i^{-1} &= \begin{bmatrix} 0 & -1 \\ 1 & 0 \end{bmatrix} \\
z\left(-i^{-1}\right) &= \begin{bmatrix} 2 & -3 \\ 3 & 2 \end{bmatrix} \begin{bmatrix} 0 & -1 \\ 1 & 0 \end{bmatrix} = \begin{bmatrix} -3 & -2 \\ 2 & -3 \end{bmatrix}.
\end{aligned}
$$

3.9.9 *The Inverse of a Complex Number*

Compute the inverse of $2 + 3i$.

Complex Number

$$
\begin{aligned}
\frac{1}{2+3i} &= \frac{1}{(2+3i)} \frac{(2-3i)}{(2-3i)} \\
&= \frac{2-3i}{13} \\
&= \tfrac{2}{13} - \tfrac{3}{13} i.
\end{aligned}
$$

Ordered Pair

$$
\begin{aligned}
\frac{1}{(2,\ 3)} &= \frac{1}{(2,\ 3)} \frac{(2,\ -3)}{(2,\ -3)} \\
&= \frac{(2,\ -3)}{13} \\
&= \left(\tfrac{2}{13},\ -\tfrac{3}{13}\right).
\end{aligned}
$$

Matrix

$$
\begin{aligned}
z &= \begin{bmatrix} 2 & -3 \\ 3 & 2 \end{bmatrix} \\
z^{-1} &= \tfrac{1}{13} \begin{bmatrix} 2 & 3 \\ -3 & 2 \end{bmatrix}.
\end{aligned}
$$

3.9.10 *The Inverse of i*

Compute the inverse of i.

Complex Number

$$\frac{1}{0+i} = \frac{1}{(0+i)}\frac{(0-i)}{(0-i)} = \frac{-i}{1} = -i.$$

Ordered Pair

$$\frac{1}{(0,\ 1)} = \frac{1}{(0,\ 1)}\frac{(0,\ -1)}{(0,\ -1)} = \frac{(0,\ -1)}{(1,\ 0)} = (0,\ -1) = -i.$$

Matrix

$$i = \begin{bmatrix} 0 & -1 \\ 1 & 0 \end{bmatrix}$$

$$i^{-1} = \begin{bmatrix} 0 & 1 \\ -1 & 0 \end{bmatrix} = -i.$$

3.9.11 *The Inverse of −i*

Compute the inverse of $-i$.

Complex Number

$$\frac{1}{0-i} = \frac{1}{(0-i)}\frac{(0+i)}{(0+i)} = \frac{i}{1} = i.$$

Ordered Pair

$$\frac{1}{(0,\ -1)} = \frac{1}{(0,\ -1)}\frac{(0,\ 1)}{(0,\ 1)} = \frac{(0,\ 1)}{(1,\ 0)} = (0,\ 1) = i.$$

Matrix

$$-i = \begin{bmatrix} 0 & 1 \\ -1 & 0 \end{bmatrix}$$

$$-i^{-1} = \begin{bmatrix} 0 & -1 \\ 1 & 0 \end{bmatrix} = i.$$

References

1. Vince, J.: Imaginary Mathematics for Computer Science. Springer, Berlin (2018). ISBN 978-3-319-94636-8
2. Descartes, R.: La Géométrie (1637) (There is an English translation by Michael Mahoney) Dover, New York (1979)

Chapter 4
The Complex Plane

4.1 Introduction

The history of some subjects often makes exciting reading, especially when there is a dispute over dates or prior art. Clarifying who did something before someone else is the work of historians, who can help unravel who should be blamed for an event, and who should take the credit. Untangling events from journals, books and private letters, and placing them in an unbiased, temporal sequence requires subject knowledge, tenacity and objective analysis.

For most research disciplines, two dates are very important in establishing priority: the date a paper is submitted for publication, and the date an accepted paper is published. Such a protocol appears to be a fair scheme, but nevertheless, assumes an efficient postal system, an unbiased peer review system, and much else.

In mathematics and the sciences, some researchers are not always confident about releasing an embryonic idea for publication, and if not published, the idea either remains in their head, or on their desk in a notebook, which may or may not be discovered after the researcher's death. Unfortunately for the researcher, a human head is not a convenient depository of information for the historian!

Sometimes, mathematical papers appear in journals associated with other disciplines, which, understandably, are not necessarily monitored by the mathematics community. Again, clever detective work on the part of historians or inquisitive academics, bring to the surface complex issues of priority, attribution, and in some cases, the unsavoury suspicion of plagiarism.

The invention of the complex plane is a perfect example of how things can go seriously wrong for the inventor when official channels for publishing mathematical ideas are bipassed. Let's see what happened.

J. Vince, *Quaternions for Computer Graphics*,
https://doi.org/10.1007/978-1-4471-7509-4_4

4.2 Some History

It all started in 1813, when the amateur Swiss mathematician Jean-Robert Argand (1768–1822), published his idea on the geometric interpretation of complex numbers in a '*brochure*' he privately funded: *Essai sur une manière de représenter les quantités imaginaires dans les constructions géométriques* [1].

The brochure was not widely distributed, and to make matters worse, it did not carry Argand's name! In a very roundabout way, the brochure's contents were eventually discovered, and in 1813 Jacques Français republished the idea of the complex plane in a paper, and requested the anonymous author of the original idea to reveal his identity. Argand came forward and was given credit for his invention, and today the complex plane is known as the *Argand diagram.* A second edition of Argand's work was published in 1874 by the publishers Gauthier-Villars [2].

Unbeknown to Argand—and everyone else at the time—a Norwegian surveyor Caspar Wessel (1745–1818), had been triangulating Denmark and developing mathematical techniques to simplify his work. One of these ideas was the original idea of adding vectorial quantities, the other was the geometric interpretation of complex numbers.

Wessel presented his first and only mathematical paper describing his complex plane to a meeting of the Royal Danish Academy in 1797, and it was published in the Academy's *Mémoires* in 1799. Wessel's paper remained hidden from the mathematical community for almost a century, when it was discovered in 1895 by the Danish mathematician Sophus Christian Juel (1855–1935). However, although everyone now agrees that Wessel was the first person to invent the complex plane, it still bears Argand's name.

But it doesn't end there! The Scottish mathematician Peter Guthrie Tait (1831–1901), wrote in his book *An Elementary Treatise on Quaternions*:

> Wallis, in the end of the seventeenth century, proposed to represent the impossible roots, of a quadratic equation by going *out of* the line on which, if real they would have been laid off. His construction is equivalent to the consideration of $\sqrt{-1}$ as a directed unit line perpendicular to that on which real quantities are measured. [3]

John Wallis (1616–1703), was a gifted English mathematician [4], and it is believed that Argand, Warren, and others, extended the results of Wallis and De Moivre, who had done some early work on the complex plane.

4.3 The Complex Plane

One of the people associated with the development of complex numbers was the brilliant Swiss mathematician Leonhard Euler (1707–1783). Euler proved the identity:

$$e^{i\theta} = \cos\theta + i\sin\theta$$

and when $\theta = \pi$, one of the most beautiful formulae in mathematics emerges:

$$e^{i\pi} = -1$$

or

$$e^{i\pi} + 1 = 0$$

which integrates five important constants: 0, 1, e, π and i, as well as the basic arithmetic operations: addition, multiplication and exponentiation.

Another consequence of this formula arises when $\theta = \pi/2$:

$$e^{i\pi/2} = \cos\left(\tfrac{\pi}{2}\right) + i\sin\left(\tfrac{\pi}{2}\right) = i$$

therefore,

$$\begin{aligned} i^i &= \left(e^{i\pi/2}\right)^i \\ &= e^{i^2\pi/2} \\ &= e^{-\pi/2} \\ &\approx 0.207879576. \end{aligned}$$

which shows that the imaginary unit raised to itself equals a real number!

In Chap. 3 we saw that the powers of imaginary i give rise to two sequences $(1, i, -1, -i, 1, \dots)$ and $(1, -i, -1, i, 1, \dots)$ which bear a striking resemblance to the patterns $(x, y, -x, -y, x, \dots)$ and $(x, -y, -x, y, x, \dots)$ that arise when rotating about the Cartesian axes in an counter-clockwise and clockwise direction, respectively. This resemblance is no coincidence, as complex numbers belong to a two-dimensional plane called the *complex plane*, which we will now describe.

The complex plane enables us to visualise complex numbers using the horizontal axis to record the real part, and the vertical axis to record the imaginary part, as shown in Fig. 4.1.

The figure also shows a circle with unit radius passing through the points $1, i, -1, -i$, which is the sequence associated with increasing powers of i. We can see the positions for $i^0 = 1, i^1 = i, i^2 = -1, i^3 = -i$ and $i^4 = 1$, which suggest that multiplying by i is equivalent to rotating through 90°.

To demonstrate this rotational effect, Fig. 4.2 shows the complex plane with four complex numbers:

$$p = 2 + i, \quad q = -1 + 2i, \quad r = -2 - i, \quad s = 1 - 2i$$

which are 90° apart.

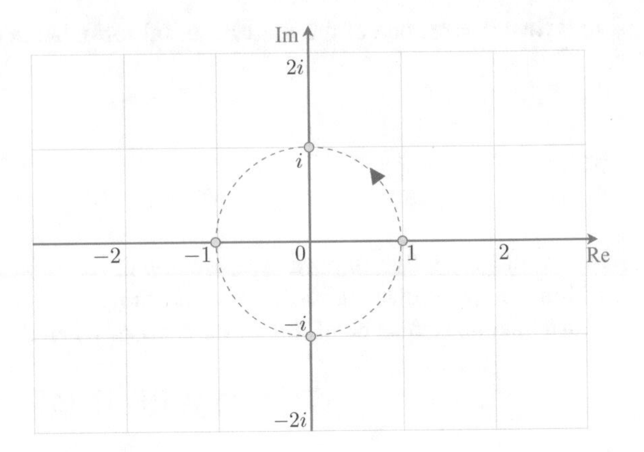

Fig. 4.1 The complex plane with the unit circle

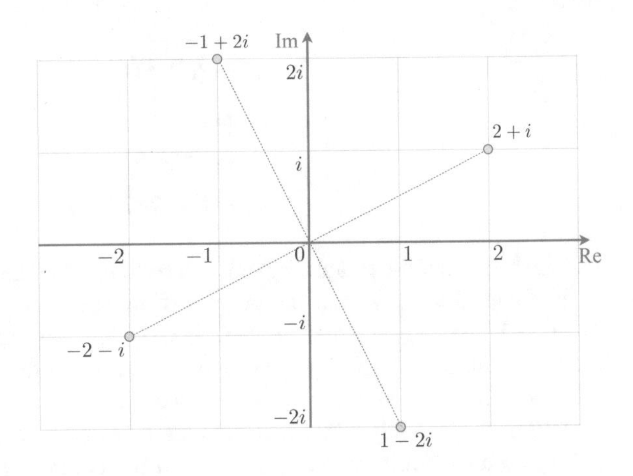

Fig. 4.2 The complex plane with four complex numbers

The point p is rotated 90° to q by multiplying it by i:

$$\begin{aligned} i(2+i) &= 2i + i^2 \\ &= -1 + 2i. \end{aligned}$$

The point q is rotated another 90° to r by multiplying it by i:

$$\begin{aligned} i(-1+2i) &= -i + 2i^2 \\ &= -2 - i. \end{aligned}$$

The point r is rotated another 90° to s by multiplying it by i:

$$\begin{aligned} i(-2-i) &= -2i - i^2 \\ &= 1 - 2i. \end{aligned}$$

Finally, the point s is rotated 90° back to p by multiplying it by i:

$$\begin{aligned} i(1-2i) &= i - 2i^2 \\ &= 2 + i. \end{aligned}$$

We also discovered in Chap. 3 that the sequence associated with increasing negative powers: $(1, -i, -1, i, \ldots)$ is a rotation in a clockwise direction, and implies that dividing a complex number by i rotates it 90° clockwise. However, we showed that $i^{-1} = -i$, and it is much easier to multiply a complex number by $-i$ than divide it by i. So let's repeat the above exercise to prove this point.

The point p is rotated $-90°$ to s by multiplying it by $-i$:

$$\begin{aligned} -i(2+i) &= -2i - i^2 \\ &= 1 - 2i. \end{aligned}$$

The point s is rotated another $-90°$ to r by multiplying it by $-i$:

$$\begin{aligned} -i(1-2i) &= -i + 2i^2 \\ &= -2 - i. \end{aligned}$$

The point r is rotated another 90° to q by multiplying it by $-i$:

$$\begin{aligned} -i(-2-i) &= 2i + i^2 \\ &= -1 + 2i. \end{aligned}$$

Finally, the point q is rotated 90° back to p by multiplying it by $-i$:

$$\begin{aligned} -i(-1+2i) &= i - 2i^2 \\ &= 2 + i. \end{aligned}$$

Thus a complex number is rotated $\pm 90°$ by multiplying it by $\pm i$.

In Chap. 3 we saw that the roots of $\sqrt{\pm i}$ are

$$\begin{aligned} \sqrt{+i} &= \pm\tfrac{\sqrt{2}}{2}(1+i) \\ \sqrt{-i} &= \pm\tfrac{\sqrt{2}}{2}(1-i) \end{aligned}$$

and are shown in Fig. 4.3. Note that the individual roots are 180° apart, which suggests that angles have something to do with their action. For example, the positive root of $\sqrt{i}$ is $\sqrt{2}/2(1+i)$ and is 45° from the real axis. Multiplying this root by itself rotates it 45° to the i axis. Similarly, the negative root is $-\sqrt{2}/2(1+i)$ and is 225° from the real axis. Multiplying this root by itself rotates it 225° to the i axis. The same is true for the roots of $\sqrt{-i}$.

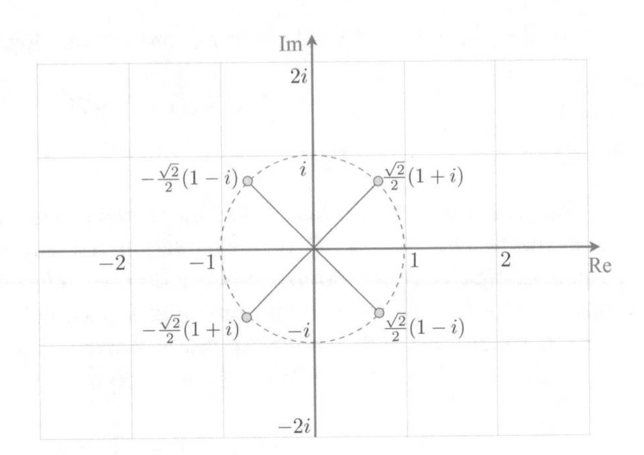

Fig. 4.3 The complex roots of $\sqrt{\pm i}$

These observations seem to suggest that we can construct a complex number capable of rotating another complex number through any angle. Which is true and is covered next.

4.4 Polar Representation

Placing a complex number on the complex plane leads us to *polar representation* where we form a line from the origin to the complex number as shown in Fig. 4.4. The length of the line is r and equals $\sqrt{a^2+b^2}$, which is why the norm of a complex number is defined using the Pythagorean formula:

$$r = |z| = \sqrt{a^2 + b^2}.$$

The angle θ between the line and the real axis is called the *argument* of z and written:

$$\arg(z) = \theta$$

where,

$$\tan\theta = \frac{b}{a}.$$

$$\text{1st quadrant: } a > 0,\ b > 0, \qquad \theta = \arctan\left(\frac{b}{a}\right).$$

$$\text{2nd \& 3rd quadrant: } a < 0, \qquad \theta = \arctan\left(\frac{b}{a}\right) + \pi.$$

$$\text{4th quadrant: } a > 0,\ b < 0, \qquad \theta = \arctan\left(\frac{b}{a}\right) + 2\pi.$$

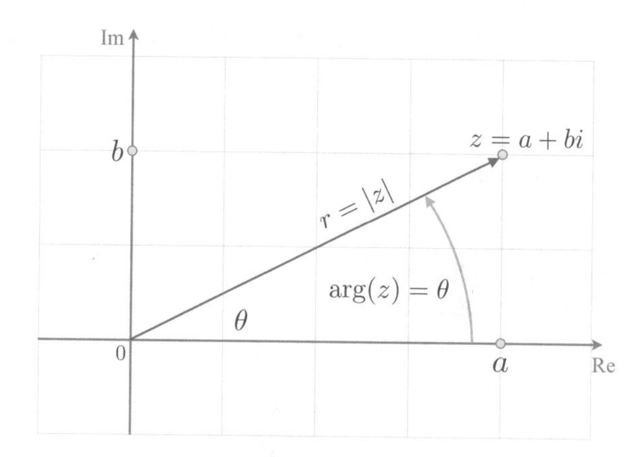

Fig. 4.4 Polar representation of a complex number

We can see from Fig. 4.4 that the horizontal component of z is $r\cos\theta$ and the vertical component is $r\sin\theta$, which permits us to write

$$\begin{aligned} z &= a + bi \\ &= r\cos\theta + ri\sin\theta \\ &= r(\cos\theta + i\sin\theta). \end{aligned}$$

As mentioned above, one of Euler's discoveries is the identity relating the power series for e^{θ}, $\sin\theta$ and $\cos\theta$:

$$e^{i\theta} = \cos\theta + i\sin\theta$$

which permits us to write

$$z = re^{i\theta}.$$

Armed with this discovery, we are now in a position to revisit the product and quotient of two complex numbers using polar representation. For example, given the following complex numbers:

$$\begin{aligned} z &= re^{i\theta} \\ w &= se^{i\phi} \end{aligned}$$

their product is

$$\begin{aligned} zw &= rse^{i\theta}e^{i\phi} \\ &= rse^{i(\theta+\phi)} \\ &= rs\big[\cos(\theta+\phi) + i\sin(\theta+\phi)\big]. \end{aligned}$$

So the product of two complex numbers creates a third one with norm

$$|zw| = rs$$

and argument

$$\arg(zw) = \theta + \phi$$

where the angles are added.

Next, the quotient:

$$\begin{aligned}\frac{z}{w} &= \frac{re^{i\theta}}{se^{i\phi}}\\ &= \frac{r}{s}e^{i(\theta-\phi)}\\ &= \frac{r}{s}\big[\cos(\theta-\phi) + i\sin(\theta-\phi)\big]\end{aligned}$$

where the norm is

$$\left|\frac{z}{w}\right| = \frac{r}{s}$$

and the argument is

$$\arg\left(\frac{z}{w}\right) = \theta - \phi$$

where the angles are subtracted.

Let's employ these formulae with an example. Figure 4.5 shows two complex numbers:

$$\begin{aligned}z &= 2 + 2i\\ w &= -1 + i\end{aligned}$$

which in polar form are

$$\begin{aligned}z &= 2\sqrt{2}\,(\cos 45^\circ + i\sin 45^\circ) = 2\sqrt{2}e^{i\pi/4}\\ w &= \sqrt{2}\,(\cos 135^\circ + i\sin 135^\circ) = \sqrt{2}e^{i3\pi/4}.\end{aligned}$$

Using normal complex algebra, the product zw is

$$zw = (2 + 2i)(-1 + i) = -4$$

and using polar form:

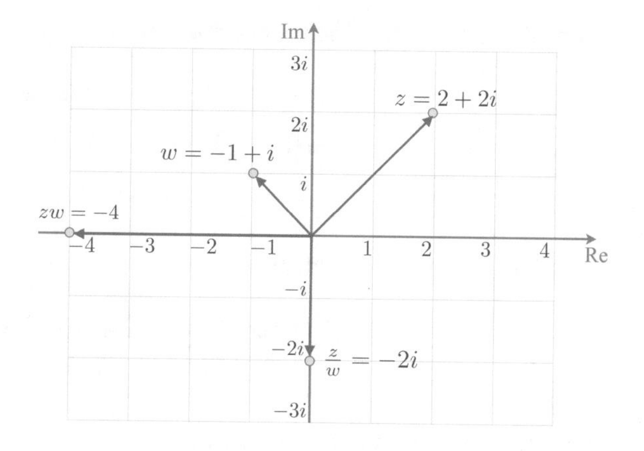

Fig. 4.5 The product and quotient of two complex numbers

$$
\begin{aligned}
|zw| &= 2\sqrt{2}\sqrt{2} = 4 \\
\arg(zw) &= 45^\circ + 135^\circ = 180^\circ
\end{aligned}
$$

which encode -4.

Now let's compute the quotient z/w using normal complex algebra, then the polar form.

$$
\begin{aligned}
\frac{z}{w} &= \frac{(2+2i)}{(-1+i)}\frac{(-1-i)}{(-1-i)} \\
&= \frac{-2-2i-2i-2i^2}{1+1} \\
&= -2i.
\end{aligned}
$$

Next, using polar form:

$$
\begin{aligned}
\left|\frac{z}{w}\right| &= \frac{2\sqrt{2}}{\sqrt{2}} = 2 \\
\arg\left(\frac{z}{w}\right) &= 45^\circ - 135^\circ = -90^\circ
\end{aligned}
$$

which encode the complex number $-2i$. These results are shown in Fig. 4.5.

We can also use Euler's formula to compute $\sqrt{i}$ as follows:

$$
e^{i\theta} = \cos\theta + i\sin\theta
$$

substituting $\theta = \pi/2$ we have

$$
e^{i\pi/2} = \cos\left(\tfrac{\pi}{2}\right) + i\sin\left(\tfrac{\pi}{2}\right) = i
$$

taking the square root of both sides, we have

$$\begin{aligned}
\pm e^{i\pi/4} &= \sqrt{i} \\
\pm\left[\cos\left(\tfrac{\pi}{4}\right) + i\sin\left(\tfrac{\pi}{4}\right)\right] &= \sqrt{i} \\
\pm\tfrac{\sqrt{2}}{2}(1+i) &= \sqrt{i}.
\end{aligned}$$

To find $\sqrt{-i}$ we substitute $\theta = -\pi/2$:

$$\begin{aligned}
e^{-i\pi/2} &= \cos\left(-\tfrac{\pi}{2}\right) + i\sin\left(-\tfrac{\pi}{2}\right) = -i \\
&= \cos\left(\tfrac{\pi}{2}\right) - i\sin\left(\tfrac{\pi}{2}\right) = -i
\end{aligned}$$

taking the square root of both sides, we have

$$\begin{aligned}
\pm e^{-i\pi/4} &= \sqrt{-i} \\
\pm\left[\cos\left(\tfrac{\pi}{4}\right) - i\sin\left(\tfrac{\pi}{4}\right)\right] &= \sqrt{-i} \\
\pm\tfrac{\sqrt{2}}{2}(1-i) &= \sqrt{-i}.
\end{aligned}$$

Higher roots can be found using a similar technique.

4.5 Rotors

The polar form brings home the fact that multiplying $z = re^{i\theta}$ with norm r, by $w = se^{i\phi}$ with norm s, creates a third complex number with norm rs. Therefore, to avoid scaling z, w must have a norm of unity. Under such conditions, w acts as a *rotor*. For example, multiplying $4+5i$ by $1+0i$ leaves it unscaled and unrotated. However, multiplying $4+5i$ by $0+i$ rotates it by 90° without any scaling.

Therefore, to rotate $2+2i$ by 45°, we must multiply it by $e^{i\pi/4}$:

$$\begin{aligned}
e^{i\pi/4} &= \cos 45^\circ + i\sin 45^\circ = \tfrac{\sqrt{2}}{2}(1+i) \\
\tfrac{\sqrt{2}}{2}(1+i)(2+2i) &= \tfrac{\sqrt{2}}{2}4i \\
&= 2\sqrt{2}i.
\end{aligned}$$

So $e^{i\theta}$ rotates any complex number through an angle θ.

To rotate a complex number $x + yi$ through an angle θ we can multiply it by the rotor $\cos\theta + i\sin\theta$:

$$\begin{aligned}
x' + y'i &= (\cos\theta + i\sin\theta)(x + yi) \\
&= x\cos\theta - y\sin\theta + i(x\sin\theta + y\cos\theta)
\end{aligned}$$

which in matrix form is:

$$\begin{bmatrix} x' \\ iy' \end{bmatrix} = \begin{bmatrix} \cos\theta & -\sin\theta \\ \sin\theta & \cos\theta \end{bmatrix} \begin{bmatrix} x \\ iy \end{bmatrix}.$$

Before moving on let's consider the effect the complex conjugate of a rotor has on rotational direction, and we can do this by multiplying $x + yi$ by the rotor $\cos\theta - i\sin\theta$:

$$\begin{aligned} x' + y'i &= (\cos\theta - i\sin\theta)(x + yi) \\ &= x\cos\theta + y\sin\theta + i(-x\sin\theta + y\cos\theta) \end{aligned}$$

and in matrix form is

$$\begin{bmatrix} x' \\ iy' \end{bmatrix} = \begin{bmatrix} \cos\theta & \sin\theta \\ -\sin\theta & \cos\theta \end{bmatrix} \begin{bmatrix} x \\ iy \end{bmatrix}$$

which is a rotation of $-\theta$ about the origin.

Therefore, we define a rotor $\mathbf{R}_\theta$ and its conjugate $\mathbf{R}_\theta^\dagger$ as

$$\begin{aligned} \mathbf{R}_\theta &= \cos\theta + i\sin\theta \\ \mathbf{R}_\theta^\dagger &= \cos\theta - i\sin\theta \end{aligned}$$

where $\mathbf{R}_\theta$ rotates $+\theta$, and $\mathbf{R}_\theta^\dagger$ rotates $-\theta$. Note the use of the dagger † symbol.

4.6 Summary

In this chapter we have discovered a graphical interpretation for complex numbers using the complex plane. Euler's formula $e^{i\theta} = \cos\theta + i\sin\theta$ permits us to express a complex number as an imaginary power of e, which in turn allows us to compute products and quotients easily. Collectively, these ideas have lead us towards the idea of a rotor, which will be developed using quaternions.

4.6.1 Summary of Definitions

Complex Number

$$\begin{aligned} z &= a + bi \\ |z| &= \sqrt{a^2 + b^2}. \end{aligned}$$

Polar Form

$$\begin{aligned}
z &= re^{i\theta} \\
z &= r(\cos\theta + i\sin\theta) \\
r &= |z| \\
\tan\theta &= b/a \\
\theta &= \arg(z).
\end{aligned}$$

$$\begin{aligned}
\text{1st quadrant:} \quad & a > 0,\ b > 0, & \theta &= \arctan\left(\frac{b}{a}\right). \\
\text{2nd \& 3rd quadrant:} \quad & a < 0, & \theta &= \arctan\left(\frac{b}{a}\right) + \pi. \\
\text{4th quadrant:} \quad & a > 0,\ b < 0, & \theta &= \arctan\left(\frac{b}{a}\right) + 2\pi.
\end{aligned}$$

Product

$$\begin{aligned}
z &= re^{i\theta} \\
w &= se^{i\phi} \\
zw &= rse^{i(\theta+\phi)} \\
&= rs\left[\cos(\theta+\phi) + i\sin(\theta+\phi)\right].
\end{aligned}$$

Quotient

$$\begin{aligned}
\frac{z}{w} &= \frac{r}{s}e^{i(\theta-\phi)} \\
&= \frac{r}{s}\left[\cos(\theta-\phi) + i\sin(\theta-\phi)\right].
\end{aligned}$$

Rotors

$$\begin{aligned}
\mathbf{R}_\theta &= \cos\theta + i\sin\theta \\
\mathbf{R}_\theta^\dagger &= \cos\theta - i\sin\theta.
\end{aligned}$$

4.7 Worked Examples

Here are some further worked examples that employ the ideas described above. In some cases, a test is included to confirm the result.

4.7.1 Rotate a Complex Number by i

Starting with $1 + 2i$, multiply the resulting complex number by i four times, and plot the result on the complex plane.

The point p is rotated 90° to q by multiplying it by i:

$$\begin{aligned} i(1 + 2i) &= i + 2i^2 \\ &= -2 + i. \end{aligned}$$

The point q is rotated another 90° to r by multiplying it by i:

$$\begin{aligned} i(-2 + i) &= -2i + i^2 \\ &= -1 - 2i. \end{aligned}$$

The point r is rotated another 90° to s by multiplying it by i:

$$\begin{aligned} i(-1 - 2i) &= -i - 2i^2 \\ &= 2 - i. \end{aligned}$$

Finally, the point s is rotated 90° back to p by multiplying it by i:

$$\begin{aligned} i(2 - i) &= 2i - i^2 \\ &= 2 + i. \end{aligned}$$

Figure 4.6 shows the four complex numbers separated by 90°.

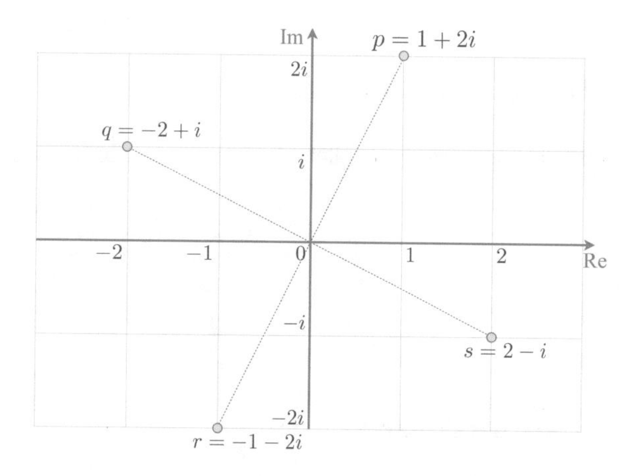

Fig. 4.6 The complex plane with four complex numbers

4.7.2 Product and Quotient Using Polar Form

Compute the product zw and quotient z/w using polar form.

$$\begin{aligned} z &= 3 + 3i \\ w &= -1 - i. \end{aligned}$$

Product:

$$\begin{aligned} z &= 3\sqrt{2}(\cos 45^\circ + i \sin 45^\circ) = 3\sqrt{2}e^{i\pi/4} \\ w &= \sqrt{2}(\cos 225^\circ + i \sin 225^\circ) = \sqrt{2}e^{i5\pi/4} \\ |zw| &= 3\sqrt{2}\sqrt{2} = 6 \\ \arg(zw) &= 45^\circ + 225^\circ = 270^\circ \end{aligned}$$

which encode the complex number $-6i$.

Test: Using normal complex algebra, the product zw is

$$zw = (3 + 3i)(-1 - i) = -6i.$$

Quotient:

$$\begin{aligned} |z| &= 3\sqrt{2} \\ |w| &= \sqrt{2} \\ \left|\frac{z}{w}\right| &= 3\sqrt{2}/\sqrt{2} = 3 \\ \arg\left(\frac{z}{w}\right) &= 45^\circ - 225^\circ = 180^\circ \end{aligned}$$

which encode the complex number -3.

Test: Using normal complex algebra, the quotient z/w is

$$\begin{aligned} \frac{z}{w} &= \frac{(3 + 3i)}{(-1 - i)}\frac{(-1 + i)}{(-1 + i)} \\ &= \frac{-6}{2} \\ &= -3 \end{aligned}$$

and agrees with the polar form.

4.7.3 Design a Rotor to Rotate a Complex Number 30°

Design a rotor to rotate a complex number through 30° without scaling.
Starting with

$$e^{i\theta} = \cos\theta + i\sin\theta$$

let $\theta = 30^\circ = \pi/6$

$$\begin{aligned} e^{i\pi/6} &= \cos 30^\circ + i\sin 30^\circ \\ &= \tfrac{\sqrt{3}}{2} + \tfrac{1}{2}i \\ &= \tfrac{1}{2}\left(\sqrt{3}+i\right). \end{aligned}$$

Test: Let's rotate $1 + 0i$ three times by this rotor to i.

$$\begin{aligned} \tfrac{1}{2}\left(\sqrt{3}+i\right)\tfrac{1}{2}\left(\sqrt{3}+i\right)\tfrac{1}{2}\left(\sqrt{3}+i\right)1 &= \tfrac{1}{8}\left(\sqrt{3}+i\right)\left(\sqrt{3}+i\right)\left(\sqrt{3}+i\right) \\ &= \tfrac{1}{8}\left(2+2\sqrt{3}i\right)\left(\sqrt{3}+i\right) \\ &= \tfrac{1}{8}\left(2\sqrt{3}-2\sqrt{3}+2i+6i\right) \\ &= i. \end{aligned}$$

4.7.4 Design a Rotor to Rotate a Complex Number −60°

Design a rotor to rotate a complex number through -60° without scaling.
Starting with

$$e^{i\theta} = \cos\theta + i\sin\theta$$

let $\theta = -60^\circ = -\pi/3$

$$\begin{aligned} e^{-i\pi/3} &= \cos(-60^\circ) + i\sin(-60^\circ) \\ &= \tfrac{1}{2} - \tfrac{\sqrt{3}}{2}i \\ &= \tfrac{1}{2}\left(1-\sqrt{3}i\right). \end{aligned}$$

References

1. Argand, J.R.: http://www-history.mcs.st-andrews.ac.uk/Mathematicians/Argand.html
2. Argand, J.R.: Essai sur une manière de représenter les quantités imaginaires dans les constructions géométriques, 2nd edn. Gauthier-Villars, Paris (1874)
3. Tait, P.G.: Elementary Treatise on Quaternions. Cambridge University Press, Cambridge (1867)
4. Wallis, J.: http://www-history.mcs.st-andrews.ac.uk/Mathematicians/Wallis.html

Chapter 5
Triples and Quaternions

5.1 Introduction

This chapter covers the period up to Hamilton's invention of quaternions. It not only shows what other mathematicians were thinking, but what Hamilton was wrestling with.

5.2 Some History

When a group of brilliant mathematicians is interested in the same subject, it is not uncommon to discover two of them coming up with the same invention at the same time. Even though two such individuals will have different mathematical strengths, they should have access to the same edifice of accumulated mathematical knowledge, and be aware of problems that have been solved, and those waiting for a solution.

We saw in Chap. 4 how Wessel and Argand had both invented the complex plane, and used it to visualise complex numbers. It was unfortunate for both men that they didn't have access to today's ubiquitous publishing network, and the Internet. Nevertheless, priority was—and still is—decided by who gets to the printing presses first. But as we saw with Wessel, even being first into print didn't guarantee fame.

A similar story surrounds the invention of quaternions. Sir William Rowan Hamilton is recognised as the inventor of quaternion algebra, which became the first non-commutative algebra to be discovered. One can imagine the elation he felt when finding a solution to a problem he had been thinking about for over a decade!

The invention provided the first mathematical framework for manipulating vectorial quantities, although this was to be refined by the American theoretical physicist, chemist, and mathematician Josiah Willard Gibbs (1839–1903). Hamilton had arrived at his invention through an algebraic and geometric route, as it was obvious to him that quaternions had significant geometric potential. Consequently, he imme-

J. Vince, *Quaternions for Computer Graphics*,
https://doi.org/10.1007/978-1-4471-7509-4_5

diately started to explore how quaternions could be applied to physics, as well as their vectorial and rotational properties.

Unbeknown to Hamilton—and virtually everyone else at the time—the French social reformer, and brilliant recreational mathematician Benjamin Olinde Rodrigues (1795–1851), had already published a paper in 1840 describing how to represent two successive rotations about different axes, by a single rotation about a third axis [1]. What is more, Rodrigues expressed his solution using a scalar and a 3-D axis, which pre-empted Hamilton's own approach using a scalar and a vector, by three years!

Simon Altmann has probably done more than any other person to set this record straight, and has published his views widely [2–5]. However, for the moment, let's continue with Hamilton's algebra and return to its rotational properties and Dr. Rodrigues later on.

The very existence of complex numbers presented a tantalising question for mathematicians of the 18th and 19th centuries. Could there be a 3-D equivalent? The answer to this question was not obvious, and many gifted mathematicians, including Gauss, Möbius, Grassmann, and Hamilton had been searching for the answer.

Hamilton's research is well documented and covers a period from the early 1830s to 1843, when he invented quaternions. And for a further 22 years, until his death in 1865, he was preoccupied with the subject. By 1833 he had shown that complex numbers form an algebra of couples, i.e. ordered pairs [6].

5.3 Triples

Hamilton recorded his thoughts in a notebook, where he outlined the events leading up to his discovery. One of the entries is:

> I, this morning, was led to what seems to me a theory of *quaternions*, which may have interesting developments. *Couples* being supposed known, and known to be representable by points in a plane, so that $\sqrt{-1}$ is perpendicular to 1, it is natural to conceive that there may be another sort of $\sqrt{-1}$, perpendicular to the plane itself. Let this new imaginary be j: so that $j^2 = -1$. A point $x,\ y,\ z$ in space may suggest the triple $x + iy + jz$. [7]

This entry shows Hamilton's thinking, that as a 2-D complex number is represented by $x + iy$, a 3-D complex number could be represented by the triple: $x + iy + jz$, where i and j are imaginary quantities that square to -1. However, the square of such a triple raises a problem with its algebraic expansion:

$$\begin{aligned} z &= x + iy + jz \\ z^2 &= (x + iy + jz)(x + iy + jz) \\ &= x^2 + ixy + jxz + ixy - y^2 + ijyz + jxz + jiyz - z^2 \\ &= x^2 - y^2 - z^2 + 2ixy + 2jxz + 2ijyz. \end{aligned} \tag{5.1}$$

The squaring operation almost closes—apart from the term $2ijyz$.

5.3.1 *Adding and Subtracting Triples*

Although Hamilton's triples refuse to close when squared, they are readily added or subtracted:

$$\begin{aligned} z_1 &= a_1 + b_1 i + c_1 j \\ z_2 &= a_2 + b_2 i + c_2 j \\ z_1 \pm z_2 &= a_1 \pm a_2 + (b_1 \pm b_2)i + (c_1 \pm c_2)j. \end{aligned}$$

Hamilton wrote to his son Archibald:

> Every morning in the early part of October 1843, on my coming down to breakfast, your brother William Edwin and yourself used to ask me: "Well, Papa, can you multiply triples?" Whereto I was always obliged to reply, with a sad shake of the head, "No, I can only add and subtract them." [8]

The problem with triples arises with their squaring as noted in (5.1), which is what to do with the $2ijyz$ term. Hamilton's notebook reflects this thinking:

> The square of this triplet $[x + iy + jz]$ is on the one hand $x^2 - y^2 - z^2 + 2ixy + 2jxz + 2ijyz$; such at least it seemed to me at first, because I assumed $ij = ji$. On the other hand, if this is to represent the third proportional to 1, 0, 0 and $x, \ y, \ z$, considered as *indicators of lines*, (namely the lines which end in the points having these coordinates, while they begin at the origin) and if this third proportional be supposed to have its length a third proportional to 1 and $\sqrt{(x^2 + y^2 + z^2)}$, and its distance twice as far removed from $1, 0, 0$ as $x, \ y, \ z$; then its real part ought to be $x^2 - y^2 - z^2$ and its two imaginary parts ought to have for coefficients $2xy$ and $2xz$; thus the term $2ijyz$ appeared de trop, and I was led to assume at first $ij = 0$. However I saw that this difficulty would be removed by supposing that $ji = -ij$. [9]

5.4 The Birth of Quaternions

The non-closure of triples presented a real problem for Hamilton, for he had toiled for over a decade trying to resolve it. Hamilton was trying to interpret the product of two line segments and in a stroke of genius he introduced a third imaginary term k, such that $k = ij$. The British sociologist, philosopher and historian of science Andrew Pickering notes:

> The introduction of the new imaginary k, defined as the product of i and j, thus constituted a further accommodation by Hamilton to an emergent resistance in thinking about the product of two arbitrary triples in terms of the algebraic and geometrical representations at once, and one aspect of this particular accommodation is worth emphasising. It amounted to a drastic shift of bridgehead in both systems of representation. More precisely, it consisted in defining a new bridgehead leading from two-place representations of complex algebra to not three–but four-place systems – the systems that quickly called *quaternions*. [10]

The introduction of a third imaginary term was only part of the solution; it was the algebraic rules controlling i, j and k that defied resolution. Finally, on 16 October, 1843, whilst walking with his wife, Lady Hamilton, along the Royal Canal in Ireland to preside at a meeting of the Royal Irish Academy, a flash of inspiration came to Hamilton, where he saw how the three imaginary terms i, j and k could resolve all their products. [11]

The solution was $z = a + bi + cj + dk$ where i, j, k all square to -1. And because of the four terms, Hamilton gave the name *quaternion*. Hamilton took the opportunity to record the event in stone, by carving the rules into the wall of Broome bridge, which he was passing at the time. Although his original inscription has not withstood years of Irish weather, a more permanent plaque now replaces it.

When Hamilton invented quaternions, he also created all sorts of names such as *tensor*, *versor* and *vector* to describe their attributes. As the inventor, it was Hamilton's prerogative to choose whatever names he wanted, and at the time, such names were associated with the notation of the period. For example, he called the quaternion's real part a *scalar*, and the imaginary part a *vector*. However today, a vector does not have any imaginary associations, which has slightly confused how quaternions are interpreted.

Simon Altmann has been very aware of these issues and helped clarify this confusion by subjecting quaternion algebra to close scrutiny, that, hitherto, was lacking. This algebraic rigour employs the idea of ordered pairs, which are easy to understand, and reveal the close relationship between quaternions and complex numbers.

Let's examine the algebra of quaternions which form the set $\mathbb{H}$ in recognition of Hamilton's achievement.

References

1. Cheng, H., Gupta, K.C.: An historical note on finite rotations. Trans. ASME J. Appl. Mech. **56**(1), 139–145 (1989)
2. Altmann, S.L.: Rotations, Quaternions and Double Groups, Dover, (2005), p. 16, ISBN-13: 978-0-486-44518-2 (1986)
3. Altmann, S.L.: Rodrigues, and the quaternion scandal. Math. Mag. **62**(5), 291–308 (1989)
4. Altmann, S.L.: Icons and Symmetries. Clarendon Press, Oxford (1992)
5. Altmann, S.L., Ortiz, E.L. (eds.): Mathematics and Social Utopias in France: Olinde Rodrigues and his Times, History of Mathematics, vol. 28. Am. Math. Soc, Providence (2005). 10: 0-8218-3860-1, ISBN-13: 978-0-8218-3860-0
6. Hamilton, W.R.: The Mathematical Papers of Sir William Rowan Hamilton, vol. I, Geometrical Optics; Conway, A.W., McDonnell, A.J. (eds.) vol. II, Dynamics; Halberstam, H., Ingram, R.E. (eds.) vol. III, Algebra, Cambridge University Press, Cambridge, (1931, 1940, 1967) (1833)
7. Hersh, R. (ed): 18 Unconventional Essays on the Nature of Mathematics, Pickering, A.: Concepts and the Mangle of Practice Constructing Quaternions, p. 263. Springer, Berlin (2006). ISBN 978 0-387-25717-9
8. Crowe, M.J.: A History of Vector Analysis. Dover, New York (1994)

9. Hersh, R. (ed): 18 Unconventional Essays on the Nature of Mathematics, Pickering, A.: Concepts and the Mangle of Practice Constructing Quaternions, p. 264. Springer, Berlin (2006). ISBN 978 0-387-25717-9
10. Hersh, R. (ed): 18 Unconventional Essays on the Nature of Mathematics, Pickering, A.: Concepts and the Mangle of Practice Constructing Quaternions, p. 270. Springer, Berlin (2006). ISBN 978 0-387-25717-9
11. Hamilton, W.R.: http://www-history.mcs.st-andrews.ac.uk/Mathematicians/Hamilton.html

Chapter 6
Quaternion Algebra

6.1 Introduction

This chapter contains further historical background to the invention of quaternions, and covers the evolution of quaternion algebra. I show how quaternion algebra is greatly simplified by treating a quaternion as an ordered pair, and provide examples of addition, subtraction, real, pure and unit quaternions. After defining the complex conjugate, norm, quaternion product, square and inverse, I show how a quaternion is represented by a matrix. The chapter concludes with a summary of the important definitions and several worked examples.

6.2 Some History

Hamilton defined a quaternion q, and its associated rules as

$$q = s + ia + jb + kc, \quad s, a, b, c \in \mathbb{R}$$

where,

$$i^2 = j^2 = k^2 = ijk = -1$$

$$\begin{aligned} ij &= k, \quad jk = i, \quad ki = j \\ ji &= -k, \ kj = -i, \ ik = -j \end{aligned}$$

References [1–3], but we tend to write quaternions as

$$q = s + ai + bj + ck.$$

J. Vince, *Quaternions for Computer Graphics*,
https://doi.org/10.1007/978-1-4471-7509-4_6

Observe from Hamilton's rules how the occurrence of ij is replaced by k. The extra imaginary k term is key to the cyclic patterns $ij = k$, $jk = i$, and $ki = j$, which are very similar to the cross product of two unit Cartesian vectors:

$$\mathbf{i} \times \mathbf{j} = \mathbf{k}, \quad \mathbf{j} \times \mathbf{k} = \mathbf{i}, \quad \mathbf{k} \times \mathbf{i} = \mathbf{j}.$$

In fact, this similarity is no coincidence, as Hamilton also invented the scalar and vector products. However, although quaternions provided an algebraic framework to describe vectors, one must acknowledge that vectorial quantities had been studied for many years prior to Hamilton.

Hamilton also saw that the i, j, k terms could represent three Cartesian unit vectors **i, j** and **k**, which had to possess imaginary qualities. i.e. $\mathbf{i}^2 = -1$, etc., which didn't go down well with some mathematicians and scientists who were suspicious of the need to involve so many imaginary terms.

Hamilton's motivation to search for a 3-D equivalent of complex numbers was part algebraic, and part geometric. For if a complex number is represented by a couple and is capable of rotating points on the plane by 90°, then perhaps a *triple* rotates points in space by 90°. In the end, a triple had to be replaced by a a quadruple—a quaternion.

One can regard Hamilton's rules from two perspectives. The first, is that they are an algebraic consequence of combining three imaginary terms. The second, is that they reflect an underlying geometric structure of space. The latter interpretation was adopted by P. G. Tait, and outlined in his book *An Elementary Treatise on Quaternions*. Tait's approach assumes three unit vectors **i, j, k** aligned with the x-, y-, z-axes respectively:

> The result of the multiplication of **i** into **j** or **ij** is defined to be the turning of **j** through a right angle in the plane perpendicular to **i** in the positive direction, in other words, the operation of **i** on **j** turns it round so as to make it coincide with **k**; and therefore briefly **ij** = **k**.
>
> To be consistent it is requisite to admit that if **i** instead of operating on **j** had operated on any other unit vector perpendicular to **i** in the plane yz, it would have turned it through a right-angle in the same direction, so that **ik** can be nothing else than −**j**.
>
> Extending to other unit vectors the definition which we have illustrated by referring to **i**, it is evident that **j** operating on **k** must bring it round to **i**, or **jk** = **i**. [4]

Tait's explanation is illustrated in Fig. 6.1a–d. Figure 6.1a shows the original alignment of **i, j, k**. Figure 6.1b shows the effect of turning **j** into **k**. Figure 6.1c shows the turning of **k** into **i**, and Fig. 6.1d shows the turning of **i** in to **j**.
So far, there is no mention of imaginary quantities—we just have:

$$\begin{aligned} \mathbf{ij} &= \mathbf{k}, \quad \mathbf{jk} = \mathbf{i}, \quad \mathbf{ki} = \mathbf{j} \\ \mathbf{ji} &= -\mathbf{k}, \mathbf{kj} = -\mathbf{i}, \mathbf{ik} = -\mathbf{j}. \end{aligned}$$

If we assume that these vectors obey the distributive and associative axioms of algebra, their imaginary qualities are exposed. For example:

$$\mathbf{ij} = \mathbf{k}$$

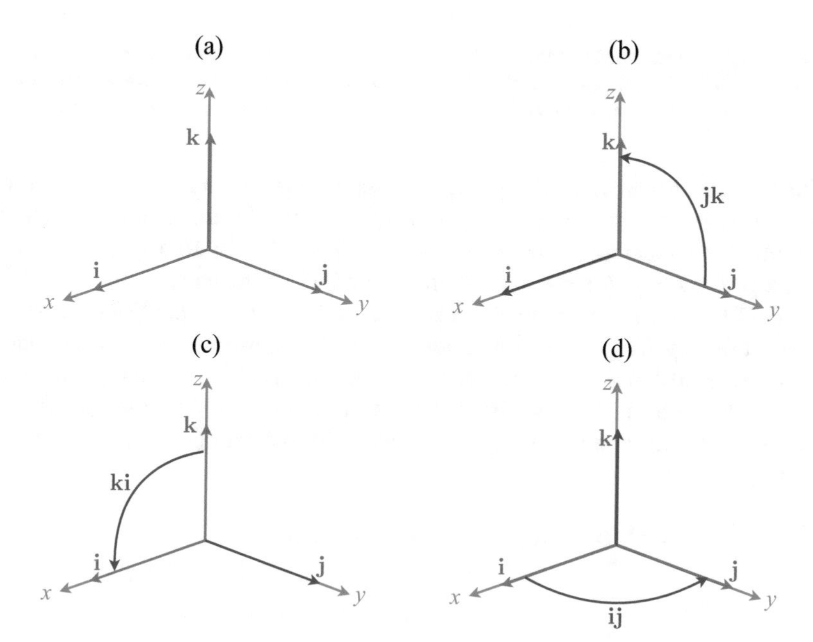

Fig. 6.1 Interpreting the products **jk**, **ki**, **ij**

and multiplying throughout by **i**:

$$\mathbf{iij} = \mathbf{ik} = -\mathbf{j}$$

therefore,

$$\mathbf{ii} = \mathbf{i}^2 = -1.$$

Similarly, we can show that $\mathbf{j}^2 = \mathbf{k}^2 = -1$.
Next:

$$\mathbf{ijk} = \mathbf{i}(\mathbf{jk}) = \mathbf{ii} = \mathbf{i}^2 = -1.$$

Thus, simply by declaring the action of the cross-product, Hamilton's rules emerge, with all of their imaginary features. Tait also made the following observation:

> A very curious speculation, due to Servois, and published in 1813 in Gergonne's Annales is the only one, so far has been discovered, in which the slightest trace of an anticipation of Quaternions is contained. Endeavouring to extend to *space* the form $a + b\sqrt{-1}$ for the plane, he is guided by analogy to write a directed unit-line in space the form
>
> $$p\cos\alpha + q\cos\beta + r\cos\gamma,$$

> where α, β, γ are its inclinations to the three axes. He perceives easily that p, q, r must be *non-reals* : but, he asks, "seraient-elles imaginaires réductibles à la forme générale $A + B\sqrt{-1}$?" This could not be the answer. In fact they are the **i**, **j**, **k** of the Quaternion Calculus. [4]

So the French mathematician François-Joseph Servois (1768–1847), was another person who came very close to discovering quaternions. Furthermore, both Tait and Hamilton were apparently unaware of the paper published by Rodrigues.

And it doesn't stop there. The brilliant German mathematician Carl Friedrich Gauss (1777–1855), was extremely cautious, and nervous of publishing anything too revolutionary, just in case he was ridiculed by fellow mathematicians. His diaries reveal that he had anticipated non-euclidean geometry ahead of Nikolai Ivanovich Lobachevsky. And in a short note from his diary in 1819 [5] he reveals that he had identified a method of finding the product of two quadruples (a, b, c, d) and $(\alpha, \beta, \gamma, \delta)$ as:

$$\begin{aligned}(A, B, C, D) &= (a,\ b,\ c,\ d)(\alpha,\ \beta,\ \gamma,\ \delta)\\ &= (a\alpha - b\beta - c\gamma - d\delta,\ a\beta + b\alpha - c\delta + d\gamma,\\ &\quad a\gamma + b\delta + c\alpha - d\beta,\ a\delta - b\gamma + c\beta + d\alpha).\end{aligned}$$

At first glance, this result does not look like a quaternion product, but if we transpose the second and third coordinates of the quadruples, and treat them as quaternions, we have:

$$\begin{aligned}(A, B, C, D) &= (a + ci + bj + dk)(\alpha + \gamma i + \beta j + \delta k)\\ &= a\alpha - c\gamma - b\beta - d\delta + a(\gamma i + \beta j + \delta k)\\ &= +\alpha(ci + bj + dk),\ (b\delta - d\beta)i + (d\gamma - c\delta)j + (c\beta - b\gamma)k\end{aligned}$$

which is identical to Hamilton's quaternion product! Furthermore, Gauss also realised that the product was non-commutative. However, he did not publish his findings, and it was left to Hamilton to invent quaternions for himself, publish his results and take the credit.

In 1881 and 1884, Josiah Willard Gibbs, at Yale University, printed his lecture notes on vector analysis for his students. Gibbs had cut the 'umbilical cord' between the real and vector parts of a quaternion and raised the 3-D vector as an independent object without any imaginary connotations. Gibbs also took on board the ideas of the German mathematician Hermann Günter Grassmann (1809–1877), who had been developing his own ideas for a vectorial system since 1832. Gibbs also defined the scalar and vector products using the relevant parts of the quaternion product. Finally, in 1901, a student of Gibbs, Edwin Bidwell Wilson, published Gibbs' notes in book form: *Vector Analysis* [6], which contains the notation in use today.

Quaternion algebra is definitely imaginary, yet simply by isolating the vector part and ignoring the imaginary rules, Gibbs was able to reveal a new branch of mathematics that exploded into vector analysis.

Hamilton and his supporters were unable to persuade their peers that quaternions could represent vectorial quantities, and eventually, Gibbs' notation won the day, and quaternions faded from the scene.

In recent years, quaternions have been rediscovered by the flight simulation industry, and more recently by the computer graphics community, where they are used to rotate vectors about an arbitrary axis. In the intervening years, various people have had the opportunity to investigate the algebra, and propose new ways of harnessing its qualities.

So let's look at three ways of annotating a quaternion q:

$$q = s + xi + yj + zk \tag{6.1}$$

$$q = s + \mathbf{v} \tag{6.2}$$

$$q = [s,\ \mathbf{v}] \tag{6.3}$$

$$\text{where} s, x, y, z \in \mathbb{R}, \quad \mathbf{v} \in \mathbb{R}^3$$

$$\text{and } i^2 = j^2 = k^2 = -1.$$

The difference is rather subtle. In (6.1) we have Hamilton's original definition with its imaginary terms and associated rules. In (6.2) a '+' sign is used to add a scalar to a vector, which seems strange, yet works. In (6.3) we have an ordered pair comprising a scalar and a vector.

Now you may be thinking: How is it possible to have three different definitions for the same object? Well, I would argue that you can call an object whatever you like, so long as they are algebraically identical. For example, matrix notation is used to represent a set of linear equations, and leads to the same results as every-day equations. Therefore, both systems of notation are equally valid.

Although I have employed the notation in (6.1) and (6.2) in other publications, in this book I have used ordered pairs. So what we need to show is that Hamilton's original definition of a quaternion (6.1), with its scalar and three imaginary terms, can be replaced by an ordered pair (6.3) comprising a scalar and a 'modern' vector.

6.3 Defining a Quaternion

Let's start with two quaternions q_a and q_b *à la* Hamilton:

$$q_a = s_a + x_a i + y_a j + z_a k$$

$$q_b = s_b + x_b i + y_b j + z_b k$$

and the obligatory rules:

$$i^2 = j^2 = k^2 = ijk = -1$$

$$ij = k, \quad jk = i, \quad ki = j$$

$$ji = -k, \; kj = -i, \; ik = -j.$$

Our objective is to show that q_a and q_b can also be represented by the ordered pairs

$$\begin{aligned} q_a &= [s_a,\ \mathbf{a}] \\ q_b &= [s_b,\ \mathbf{b}], \quad s_a, s_b \in \mathbb{R}, \quad \mathbf{a}, \mathbf{b} \in \mathbb{R}^3. \end{aligned}$$

The quaternion product $q_a q_b$ expands to

$$\begin{aligned} q_a q_b = [s_a,\ \mathbf{a}][s_b,\ \mathbf{b}] &= [s_a + x_a i + y_a j + z_a k][s_b + x_b i + y_b j + z_b k] \\ &= [(s_a s_b - x_a x_b - y_a y_b - z_a z_b) \\ &\quad + (s_a x_b + s_b x_a + y_a z_b - y_b z_a)i \\ &\quad + (s_a y_b + s_b y_a + z_a x_b - z_b x_a)j \\ &\quad + (s_a z_b + s_b z_a + x_a y_b - x_b y_a)k]. \end{aligned} \tag{6.4}$$

Equation (6.4) takes the form of another quaternion, and confirms that the quaternion product is closed.

At this stage, Hamilton turned the imaginary terms $i,\ j,\ k$ into unit Cartesian vectors $\mathbf{i},\ \mathbf{j},\ \mathbf{k}$ and transformed (6.4) into a vector form. The problem with this approach is that the vectors retain their imaginary roots. Simon Altmann's suggestion is to replace the imaginaries by the ordered pairs:

$$i = [0,\ \mathbf{i}], \quad j = [0,\ \mathbf{j}], \quad k = [0,\ \mathbf{k}]$$

which are themselves quaternions, and called *quaternion units*.

The idea of defining a quaternion in terms of quaternion units is exactly the same as defining a vector in terms of its unit Cartesian vectors. Furthermore, it permits vectors to exist without any imaginary associations.

Let's substitute these quaternion units in (6.4) together with $[1,\ \mathbf{0}] = 1$:

$$\begin{aligned} [s_a,\ \mathbf{a}][s_b,\ \mathbf{b}] &= [(s_a s_b - x_a x_b - y_a y_b - z_a z_b)[1,\ \mathbf{0}] \\ &\quad + (s_a x_b + s_b x_a + y_a z_b - y_b z_a)[0,\ \mathbf{i}] \\ &\quad + (s_a y_b + s_b y_a + z_a x_b - z_b x_a)[0,\ \mathbf{j}] \\ &\quad + (s_a z_b + s_b z_a + x_a y_b - x_b y_a)[0,\ \mathbf{k}]]. \end{aligned} \tag{6.5}$$

Next, we expand (6.5) using previously defined rules:

$$\begin{aligned} [s_a,\ \mathbf{a}][s_b,\ \mathbf{b}] &= [[s_a s_b - x_a x_b - y_a y_b - z_a z_b,\ \mathbf{0}] \\ &\quad + [0,\ (s_a x_b + s_b x_a + y_a z_b - y_b z_a)\mathbf{i}] \\ &\quad + [0,\ (s_a y_b + s_b y_a + z_a x_b - z_b x_a)\mathbf{j}] \\ &\quad + [0,\ (s_a z_b + s_b z_a + x_a y_b - x_b y_a)\mathbf{k}]]. \end{aligned} \tag{6.6}$$

A vertical scan of (6.6) reveals some hidden vectors:

$$
\begin{aligned}
[s_a,\ \mathbf{a}][s_b,\ \mathbf{b}] = [&[s_a s_b - x_a x_b - y_a y_b - z_a z_b,\ \mathbf{0}] \\
&+ [0,\ s_a(x_b\mathbf{i} + y_b\mathbf{j} + z_b\mathbf{k}) + s_b(x_a\mathbf{i} + y_a\mathbf{j} + z_a\mathbf{k}) \\
&+ (y_a z_b - y_b z_a)\mathbf{i} + (z_a x_b - z_b x_a)\mathbf{j} + (x_a y_b - x_b y_a)\mathbf{k}]].
\end{aligned} \tag{6.7}
$$

Equation (6.7) contains two ordered pairs which can now be combined:

$$
\begin{aligned}
[s_a,\ \mathbf{a}][s_b,\ \mathbf{b}] = [&s_a s_b - x_a x_b - y_a y_b - z_a z_b, \\
&+ s_a(x_b\mathbf{i} + y_b\mathbf{j} + z_b\mathbf{k}) + s_b(x_a\mathbf{i} + y_a\mathbf{j} + z_a\mathbf{k}) \\
&+ (y_a z_b - y_b z_a)\mathbf{i} + (z_a x_b - z_b x_a)\mathbf{j} + (x_a y_b - x_b y_a)\mathbf{k}].
\end{aligned} \tag{6.8}
$$

If we make

$$
\begin{aligned}
\mathbf{a} &= x_a\mathbf{i} + y_a\mathbf{j} + z_a\mathbf{k} \\
\mathbf{b} &= x_b\mathbf{i} + y_b\mathbf{j} + z_b\mathbf{k}
\end{aligned}
$$

and substitute them in (6.8) we get:

$$
[s_a,\ \mathbf{a}][s_b,\ \mathbf{b}] = [s_a s_b - \mathbf{a}\cdot\mathbf{b},\ s_a\mathbf{b} + s_b\mathbf{a} + \mathbf{a}\times\mathbf{b}] \tag{6.9}
$$

which defines the quaternion product.

From now on, we don't have to worry about Hamilton's rules as they are embedded within (6.9). Furthermore, our vectors have no imaginary associations.

Although Rodrigues did not have access to Gibbs' vector notation used in (6.9), he managed to calculate the equivalent algebraic expression, which was some achievement.

6.3.1 The Quaternion Units

Using (6.9) we can check to see if the quaternion units are imaginary by squaring them:

$$
\begin{aligned}
i &= [0,\ \mathbf{i}] \\
i^2 &= [0,\ \mathbf{i}][0,\ \mathbf{i}] \\
&= [\mathbf{i}\cdot\mathbf{i},\ \mathbf{i}\times\mathbf{i}] \\
&= [-1,\ \mathbf{0}]
\end{aligned}
$$

which is a *real quaternion* and equivalent to -1, confirming that $[0,\ \mathbf{i}]$ is imaginary. Using a similar expansion we can shown that $[0,\ \mathbf{j}]$ and $[0,\ \mathbf{k}]$ have the same property.

Now let's compute the products ij, jk and ki:

$$\begin{aligned} ij &= [0,\ \mathbf{i}][0,\ \mathbf{j}] \\ &= [-\mathbf{i}\cdot\mathbf{j},\ \mathbf{i}\times\mathbf{j}] \\ &= [0,\ \mathbf{k}] \end{aligned}$$

which is the quaternion unit k.

$$\begin{aligned} jk &= [0,\ \mathbf{j}][0,\ \mathbf{k}] \\ &= [-\mathbf{j}\cdot\mathbf{k},\ \mathbf{j}\times\mathbf{k}] \\ &= [0,\ \mathbf{i}] \end{aligned}$$

which is the quaternion unit i.

$$\begin{aligned} ki &= [0,\ \mathbf{k}][0,\ \mathbf{i}] \\ &= [-\mathbf{k}\cdot\mathbf{i},\ \mathbf{k}\times\mathbf{i}] \\ &= [0,\ \mathbf{j}] \end{aligned}$$

which is the quaternion unit j.

Next, let's confirm that $ijk = -1$:

$$\begin{aligned} ijk &= [0,\ \mathbf{i}][0,\ \mathbf{j}][0,\ \mathbf{k}] \\ &= [0,\ \mathbf{k}][0,\ \mathbf{k}] \\ &= [-\mathbf{k}\cdot\mathbf{k},\ \mathbf{k}\times\mathbf{k}] \\ &= [-1,\ \mathbf{0}] \end{aligned}$$

which is a real quaternion equivalent to -1, confirming that $ijk = -1$.

Thus the notation of ordered pairs upholds all of Hamilton's rules. However, the last double product assumes that quaternions are associative. So let's double check to show that $(ij)k = i(jk)$:

$$\begin{aligned} i(jk) &= [0,\ \mathbf{i}][0,\ \mathbf{j}][0,\ \mathbf{k}] \\ &= [0,\ \mathbf{i}][0,\ \mathbf{i}] \\ &= [-\mathbf{i}\cdot\mathbf{i},\ \mathbf{i}\times\mathbf{i}] \\ &= [-1,\ \mathbf{0}] \end{aligned}$$

which is correct.

6.3.2 Example of Quaternion Products

Although we have yet to discover how quaternions are used to rotate vectors, let's concentrate on their algebraic traits by evaluating an example.

$$q_a = [1,\ 2\mathbf{i} + 3\mathbf{j} + 4\mathbf{k}]$$
$$q_b = [2,\ 3\mathbf{i} + 4\mathbf{j} + 5\mathbf{k}]$$

the product $q_a q_b$ is

$$\begin{aligned}
q_a q_b &= [1,\ 2\mathbf{i} + 3\mathbf{j} + 4\mathbf{k}][2,\ 3\mathbf{i} + 4\mathbf{j} + 5\mathbf{k}] \\
&= [1 \times 2 - (2 \times 3 + 3 \times 4 + 4 \times 5), \\
&\quad 1(3\mathbf{i} + 4\mathbf{j} + 5\mathbf{k}) + 2(2\mathbf{i} + 3\mathbf{j} + 4\mathbf{k}) \\
&\quad + (3 \times 5 - 4 \times 4)\mathbf{i} - (2 \times 5 - 4 \times 3)\mathbf{j} + (2 \times 4 - 3 \times 3)\mathbf{k}] \\
&= [-36,\ 7\mathbf{i} + 10\mathbf{j} + 13\mathbf{k} - \mathbf{i} + 2\mathbf{j} - \mathbf{k}] \\
&= [-36,\ 6\mathbf{i} + 12\mathbf{j} + 12\mathbf{k}]
\end{aligned}$$

which is another ordered pair representing a quaternion.

Having shown that Hamilton's *imaginary* notation has a vector equivalent, and can be represented as an ordered pair, we continue with this notation and describe other features of quaternions. Note that we can abandon Hamilton's rules as they are embedded within the definition of the quaternion product, and will surface in the following definitions.

6.4 Algebraic Definition

A quaternion is the ordered pair:

$$q = [s,\ \mathbf{v}], \quad s \in \mathbb{R}, \quad \mathbf{v} \in \mathbb{R}^3.$$

If we express $\mathbf{v}$ in terms of its components, we have

$$q = [s,\ x\mathbf{i} + y\mathbf{j} + z\mathbf{k}], \quad s, x, y, z \in \mathbb{R}.$$

6.5 Adding and Subtracting Quaternions

Addition and subtraction employ the following rule:

$$q_a = [s_a,\ \mathbf{a}]$$
$$q_b = [s_b,\ \mathbf{b}]$$
$$q_a \pm q_b = [s_a \pm s_b,\ \mathbf{a} \pm \mathbf{b}].$$

For example:

$$\begin{aligned} q_a &= [0.5,\ 2\mathbf{i} + 3\mathbf{j} - 4\mathbf{k}] \\ q_b &= [0.1,\ 4\mathbf{i} + 5\mathbf{j} + 6\mathbf{k}] \\ q_a + q_b &= [0.6,\ 6\mathbf{i} + 8\mathbf{j} + 2\mathbf{k}] \\ q_a - q_b &= [0.4,\ -2\mathbf{i} - 2\mathbf{j} - 10\mathbf{k}]. \end{aligned}$$

6.6 Real Quaternion

A *real quaternion* has a zero vector term:

$$q = [s,\ \mathbf{0}].$$

The product of two real quaternions is

$$\begin{aligned} q_a &= [s_a,\ \mathbf{0}] \\ q_b &= [s_b,\ \mathbf{0}] \\ q_a q_b &= [s_a,\ \mathbf{0}][s_b,\ \mathbf{0}] \\ &= [s_a s_b,\ \mathbf{0}] \end{aligned}$$

which is another real quaternion, and shows that they behave just like real numbers:

$$[s,\ \mathbf{0}] \equiv s.$$

We have already come across this with complex numbers containing a zero imaginary term:

$$a + bi = a, \quad \text{when } b = 0.$$

6.7 Multiplying a Quaternion by a Scalar

Intuition suggests that multiplying a quaternion by a scalar should obey the rule:

$$\begin{aligned} q &= [s,\ \mathbf{v}] \\ \lambda q &= \lambda[s,\ \mathbf{v}], \quad \lambda \in \mathbb{R} \\ &= [\lambda s,\ \lambda \mathbf{v}]. \end{aligned}$$

For example:

$$\begin{aligned} q &= 3[2,\ 3\mathbf{i} + 4\mathbf{j} + 5\mathbf{k}] \\ &= [6,\ 9\mathbf{i} + 12\mathbf{j} + 15\mathbf{k}]. \end{aligned}$$

We can confirm our intuition by multiplying a quaternion by a scalar in the form of a real quaternion:

$$\begin{aligned} q &= [s,\ \mathbf{v}] \\ \lambda &= [\lambda,\ \mathbf{0}] \\ \lambda q &= [\lambda,\ \mathbf{0}][s,\ \mathbf{v}] \\ &= [\lambda s,\ \lambda \mathbf{v}] \end{aligned}$$

which is excellent confirmation.

6.8 Pure Quaternion

Hamilton defined a *pure quaternion* as one having a zero scalar term:

$$q = xi + yj + zk$$

and is just a vector, but with imaginary qualities. Simon Altmann, and others, believe that this was a serious mistake on Hamilton's part to call a quaternion with a zero real term, a vector.

The main issue is that there are two types of vectors: *polar* and *axial*, also called a *pseudovector*. Richard Feynman describes polar vectors as '*honest*' vectors [7] and represent the every-day vectors of directed lines. Whereas, axial vectors are computed from polar vectors, such as in a vector product. However, these two types of vector do not behave in the same way when transformed. For example, given two 'honest', polar vectors **a** and **b**, we can compute the axial vector: $\mathbf{c} = \mathbf{a} \times \mathbf{b}$. Next, if we subject **a** and **b** to an inversion transform through the origin, such that **a** becomes $-\mathbf{a}$, and **b** becomes $-\mathbf{b}$, and compute their cross product $(-\mathbf{a}) \times (-\mathbf{b})$, we still get **c**! Which implies that the axial vector **c** must not be transformed along with **a** and **b**.

It could be argued that the inversion transform is not a 'proper' transform as it turns a right-handed set of axes into a left-handed set. But in physics, laws of nature are expected to work in either system. Unfortunately, Hamilton was not aware of this distinction, as he had only just invented vectors. However, in the intervening years, it has become evident that Hamilton's quaternion vector is an axial vector, and not a polar vector.

As we will see, in 3-D rotations quaternions take the form

$$q = \left[\cos\left(\tfrac{\theta}{2}\right),\ \sin\left(\tfrac{\theta}{2}\right)\mathbf{v}\right]$$

where θ is the angle of rotation and $\mathbf{v}$ is the axis of rotation, and when we set $\theta = 180^\circ$, we get

$$q = [0,\ \mathbf{v}]$$

which remains a quaternion, even though it only contains a vector part.

Consequently, we define a *pure quaternion* as

$$q = [0,\ \mathbf{v}].$$

The product of two pure quaternions is

$$\begin{aligned} q_a &= [0,\ \mathbf{a}] \\ q_b &= [0,\ \mathbf{b}] \\ q_a q_b &= [0,\ \mathbf{a}][0,\ \mathbf{b}] \\ &= [-\mathbf{a}\cdot\mathbf{b},\ \mathbf{a}\times\mathbf{b}] \end{aligned}$$

which is no longer 'pure', as some of the original vector information has 'tunnelled' across into the real part via the dot product.

6.9 Unit Quaternion

Let's pursue this analysis further by introducing some familiar vector notation.

Give vector $\mathbf{v}$, then

$$\mathbf{v} = \lambda\hat{\mathbf{v}}, \quad \text{where } \lambda = \|\mathbf{v}\| \text{ and } \|\hat{\mathbf{v}}\| = 1.$$

Combining this with the definition of a pure quaternion we get:

$$\begin{aligned} q &= [0,\ \mathbf{v}] \\ &= [0,\ \lambda\hat{\mathbf{v}}] \\ &= \lambda[0,\ \hat{\mathbf{v}}] \end{aligned}$$

and reveals the object $[0,\ \hat{\mathbf{v}}]$ which is called the *unit quaternion* and comprises a zero scalar and a unit vector. It is convenient to identify this unit quaternion as $\hat{q}$:

$$\hat{q} = [0,\ \hat{\mathbf{v}}].$$

So now we have a notation similar to that of vectors where a vector $\mathbf{v}$ is described in terms of its unit form:

$$\mathbf{v} = \lambda\hat{\mathbf{v}}$$

and a quaternion q is also described in terms of its unit form:

$$q = \lambda\hat{q}.$$

Note that $\hat{q}$ is an imaginary object as it squares to -1:

$$\begin{aligned}
\hat{q}^2 &= [0,\ \hat{\mathbf{v}}][0,\ \hat{\mathbf{v}}] \\
&= [-\hat{\mathbf{v}} \cdot \hat{\mathbf{v}},\ \hat{\mathbf{v}} \times \hat{\mathbf{v}}] \\
&= [-1,\ \mathbf{0}] \\
&= -1
\end{aligned}$$

which is not too surprising, bearing in mind Hamilton's original invention!

6.10 Additive Form of a Quaternion

We now come to the idea of splitting a quaternion into its constituent parts: a real quaternion and a pure quaternion. Again, intuition suggests that we can write a quaternion as

$$\begin{aligned}
q &= [s,\ \mathbf{v}] \\
&= [s,\ \mathbf{0}] + [0,\ \mathbf{v}]
\end{aligned}$$

and we can test this by forming the algebraic product of two quaternions represented in this way:

$$\begin{aligned}
q_a &= [s_a,\ \mathbf{0}] + [0,\ \mathbf{a}] \\
q_b &= [s_b,\ \mathbf{0}] + [0,\ \mathbf{b}] \\
q_a q_b &= \big([s_a,\ \mathbf{0}] + [0,\ \mathbf{a}]\big)\big([s_b,\ \mathbf{0}] + [0,\ \mathbf{b}]\big) \\
&= [s_a,\ \mathbf{0}][s_b,\ \mathbf{0}] + [s_a,\ \mathbf{0}][0,\ \mathbf{b}] + [0,\ \mathbf{a}][s_b,\ \mathbf{0}] + [0,\ \mathbf{a}][0,\ \mathbf{b}] \\
&= [s_a s_b,\ \mathbf{0}] + [0,\ s_a\mathbf{b}] + [0,\ s_b\mathbf{a}] + [-\mathbf{a} \cdot \mathbf{b},\ \mathbf{a} \times \mathbf{b}] \\
&= [s_a s_b - \mathbf{a} \cdot \mathbf{b},\ s_a\mathbf{b} + s_b\mathbf{a} + \mathbf{a} \times \mathbf{b}]
\end{aligned}$$

which is correct, and confirms that the additive form works.

6.11 Binary Form of a Quaternion

Having shown that the additive form of a quaternion works, and discovered the unit quaternion, we can join the two objects together as follows:

$$
\begin{aligned}
q &= [s,\ \mathbf{v}] \\
&= [s,\ \mathbf{0}] + [0,\ \mathbf{v}] \\
&= [s,\ \mathbf{0}] + \lambda[0,\ \hat{\mathbf{v}}] \\
&= s + \lambda\hat{q}.
\end{aligned}
$$

Just to recap, s is a scalar, λ is the length of the vector term, and $\hat{q}$ is the unit quaternion $[0,\ \hat{\mathbf{v}}]$.

Look how similar this notation is to a complex number:

$$
\begin{aligned}
z &= a + bi \\
q &= s + \lambda\hat{q}
\end{aligned}
$$

where $a,\ b,\ s,\ \lambda$ are scalars, i is the unit imaginary and $\hat{q}$ is the unit quaternion.

6.12 The Complex Conjugate of a Quaternion

We have already discovered that the conjugate of a complex number $z = a + bi$ is given by

$$
z^* = a - bi
$$

and is very useful in computing the inverse of z. The *quaternion conjugate* plays a similar role in computing the inverse of a quaternion. Therefore, given

$$
q = [s,\ \mathbf{v}]
$$

the quaternion conjugate is defined as

$$
q^* = [s,\ -\mathbf{v}].
$$

For example:

$$
\begin{aligned}
q &= [2,\ 3\mathbf{i} - 4\mathbf{j} + 5\mathbf{k}] \\
q^* &= [2,\ -3\mathbf{i} + 4\mathbf{j} - 5\mathbf{k}]
\end{aligned}
$$

If we compute the product qq^* we obtain

$$
\begin{aligned}
qq^* &= [s,\ \mathbf{v}][s,\ -\mathbf{v}] \\
&= \left[s^2 - \mathbf{v}\cdot(-\mathbf{v}),\ -s\mathbf{v} + s\mathbf{v} + \mathbf{v}\times(-\mathbf{v})\right] \\
&= \left[s^2 + \mathbf{v}\cdot\mathbf{v},\ \mathbf{0}\right] \\
&= \left[s^2 + v^2,\ \mathbf{0}\right].
\end{aligned}
$$

Let's show that $qq^* = q^*q$:

$$
\begin{aligned}
q^*q &= [s,\ -\mathbf{v}][s,\ \mathbf{v}] \\
&= \left[s^2 - (-\mathbf{v})\cdot\mathbf{v},\ s\mathbf{v} - s\mathbf{v} + (-\mathbf{v})\times\mathbf{v}\right] \\
&= \left[s^2 + \mathbf{v}\cdot\mathbf{v},\ \mathbf{0}\right] \\
&= \left[s^2 + v^2,\ \mathbf{0}\right] \\
&= qq^*.
\end{aligned}
$$

Now let's show that $(q_a q_b)^* = q_b^* q_a^*$.

$$
\begin{aligned}
q_a &= [s_a,\ \mathbf{a}] \\
q_b &= [s_b,\ \mathbf{b}] \\
q_a q_b &= [s_a,\ \mathbf{a}][s_b,\ \mathbf{b}] \\
&= [s_a s_b - \mathbf{a}\cdot\mathbf{b},\ s_a\mathbf{b} + s_b\mathbf{a} + \mathbf{a}\times\mathbf{b}] \\
(q_a q_b)^* &= [s_a s_b - \mathbf{a}\cdot\mathbf{b},\ -s_a\mathbf{b} - s_b\mathbf{a} - \mathbf{a}\times\mathbf{b}]. \qquad (6.10)
\end{aligned}
$$

Next, we compute $q_b^* q_a^*$

$$
\begin{aligned}
q_a^* &= [s_a,\ -\mathbf{a}] \\
q_b^* &= [s_b,\ -\mathbf{b}] \\
q_b^* q_a^* &= [s_b,\ -\mathbf{b}][s_a,\ -\mathbf{a}] \\
&= [s_a s_b - \mathbf{a}\cdot\mathbf{b},\ -s_a\mathbf{b} - s_b\mathbf{a} - \mathbf{a}\times\mathbf{b}]. \qquad (6.11)
\end{aligned}
$$

And as (6.10) equals (6.11), $(q_a q_b)^* = q_b^* q_a^*$.

6.13 Norm of a Quaternion

The *norm* of a complex number $z = a + bi$ is defined as:

$$|z| = \sqrt{a^2 + b^2}$$

which allows us to write

$$zz^* = |z|^2.$$

Similarly, the norm of a quaternion q is defined as:

$$\begin{aligned} q &= [s, \ \mathbf{v}] \\ &= [s, \ \lambda\hat{\mathbf{v}}] \\ |q| &= \sqrt{s^2 + \lambda^2} \end{aligned}$$

where $\lambda = \|\mathbf{v}\|$ which allows us to write

$$qq^* = |q|^2.$$

For example:

$$\begin{aligned} q &= [1, \ 4\mathbf{i} + 4\mathbf{j} - 4\mathbf{k}] \\ |q| &= \sqrt{1^2 + 4^2 + 4^2 + (-4)^2} \\ &= \sqrt{49} \\ &= 7. \end{aligned}$$

6.14 Normalised Quaternion

A quaternion with a unit norm is called a *normalised quaternion*. For example, the quaternion $q = [s, \ \mathbf{v}]$ is *normalised* by dividing it by $|q|$:

$$q' = \frac{q}{\sqrt{s^2 + \lambda^2}}.$$

We must be careful not to confuse the unit quaternion with a unit-norm quaternion. The unit quaternion is $[0, \ \hat{\mathbf{v}}]$ with a unit-vector part, whereas a unit-norm quaternion is normalised such that $s^2 + \lambda^2 = 1$.

I will be careful to distinguish between these two terms as many authors—including myself—use the term unit quaternion to describe a quaternion with a unit norm. For example:

$$q = [1, \ 4\mathbf{i} + 4\mathbf{j} - 4\mathbf{k}]$$

has a norm of 7, and q is normalised by dividing by 7:

$$q' = \tfrac{1}{7}\,[1, \ 4\mathbf{i} + 4\mathbf{j} - 4\mathbf{k}].$$

The type of unit-norm quaternion we will be using takes the form:

$$q = \left[\cos\left(\tfrac{\theta}{2}\right), \ \sin\left(\tfrac{\theta}{2}\right)\hat{\mathbf{v}}\right]$$

because $\cos^2\theta + \sin^2\theta = 1$.

6.15 Quaternion Products

Having shown that ordered pairs can represent a quaternion and its various manifestations, let's summarise the products we will eventually encounter. To start, we have the product of two normal quaternions:

$$\begin{aligned} q_a q_b &= [s_a,\ \mathbf{a}][s_b,\ \mathbf{b}] \\ &= [s_a s_b - \mathbf{a}\cdot\mathbf{b},\ s_a\mathbf{b} + s_b\mathbf{a} + \mathbf{a}\times\mathbf{b}]. \end{aligned}$$

6.15.1 Product of Pure Quaternions

Given two pure quaternions:

$$\begin{aligned} q_a &= [0,\ \mathbf{a}] \\ q_b &= [0,\ \mathbf{b}] \end{aligned}$$

their product is

$$\begin{aligned} q_a q_b &= [0,\ \mathbf{a}][0,\ \mathbf{b}] \\ &= [-\mathbf{a}\cdot\mathbf{b},\ \mathbf{a}\times\mathbf{b}]. \end{aligned}$$

6.15.2 Product of Unit-Norm Quaternions

Given two unit-norm quaternions:

$$\begin{aligned} q_a &= [s_a,\ \mathbf{a}] \\ q_b &= [s_b,\ \mathbf{b}] \end{aligned}$$

where $|q_a| = |q_b| = 1$. Their product is another unit-norm quaternion, which is proved as follows.

We assume $q_c = [s_c,\ \mathbf{c}]$ and show that $|q_c| = s_c^2 + c^2 = 1$ where

$$\begin{aligned} [s_c,\ \mathbf{c}] &= [s_a,\ \mathbf{a}][s_b,\ \mathbf{b}] \\ &= [s_a s_b - \mathbf{a}\cdot\mathbf{b},\ s_a\mathbf{b} + s_b\mathbf{a} + \mathbf{a}\times\mathbf{b}]. \end{aligned}$$

Let's assume the angle between **a** and **b** is θ, which permits us to write:

$$\begin{aligned} s_c &= s_a s_b - ab\cos\theta \\ \mathbf{c} &= s_a b\hat{\mathbf{b}} + s_b a\hat{\mathbf{a}} + ab\sin\theta\left(\hat{\mathbf{a}}\times\hat{\mathbf{b}}\right). \end{aligned}$$

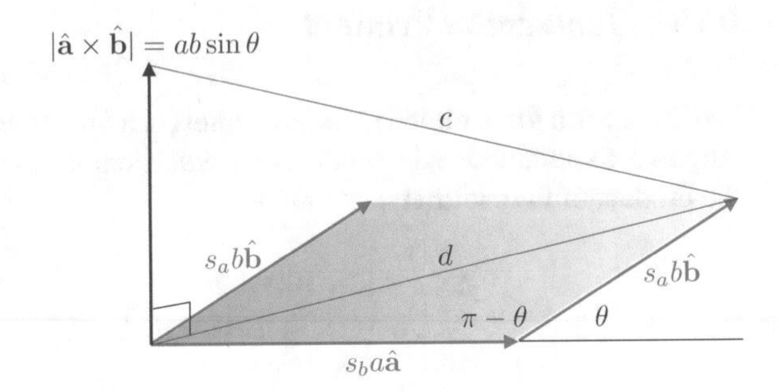

Fig. 6.2 Geometry for $s_a b\hat{\mathbf{b}} + s_b a\hat{\mathbf{a}} + ab\sin\theta(\hat{\mathbf{a}} \times \hat{\mathbf{b}})$

Therefore,

$$\begin{aligned} s_c^2 &= (s_a s_b - ab\cos\theta)(s_a s_b - ab\cos\theta) \\ &= s_a^2 s_b^2 - 2 s_a s_b ab\cos\theta + a^2 b^2 \cos^2\theta. \end{aligned}$$

Figure 6.2 shows the geometry representing **c**.

$$\begin{aligned} d^2 &= s_b^2 a^2 + s_a^2 b^2 - 2 s_a s_b ab\cos(\pi - \theta) \\ &= s_b^2 a^2 + s_a^2 b^2 + 2 s_a s_b ab\cos\theta \\ c^2 &= d^2 + a^2 b^2 \sin^2\theta \\ &= s_b^2 a^2 + s_a^2 b^2 + 2 s_a s_b ab\cos\theta + a^2 b^2 \sin^2\theta \\ s_c^2 + c^2 &= s_a^2 s_b^2 - 2 s_a s_b ab\cos\theta + a^2 b^2\cos^2\theta + s_b^2 a^2 + s_a^2 b^2 + 2 s_a s_b ab\cos\theta + a^2 b^2 \sin^2\theta \\ &= s_a^2 s_b^2 + a^2 b^2 + s_b^2 a^2 + s_a^2 b^2 \\ &= s_a^2\left(s_b^2 + b^2\right) + a^2\left(s_b^2 + b^2\right) \\ &= s_a^2 + a^2 \\ &= 1. \end{aligned}$$

Therefore, the product of two unit-norm quaternions is another unit-norm quaternion. Consequently, multiplying a quaternion by a unit-norm quaternion, does not change its norm:

$$\begin{aligned} q_a &= [s_a,\ \mathbf{a}] \\ |q_a| &= 1 \\ q_b &= [s_b,\ \mathbf{b}] \\ |q_a q_b| &= |q_b|. \end{aligned}$$

6.15.3 *Square of a Quaternion*

The square of a quaternion is given by:

$$
\begin{aligned}
\mathbf{v} &= x\mathbf{i} + y\mathbf{j} + z\mathbf{k} \\
q &= [s,\ \mathbf{v}] \\
q^2 &= [s,\ \mathbf{v}][s,\ \mathbf{v}] \\
&= \left[s^2 - \mathbf{v}\cdot\mathbf{v},\ 2s\mathbf{v} + \mathbf{v}\times\mathbf{v}\right] \\
&= \left[s^2 - \mathbf{v}\cdot\mathbf{v},\ 2s\mathbf{v}\right] \\
&= \left[s^2 - x^2 - y^2 - z^2,\ 2s(x\mathbf{i} + y\mathbf{j} + z\mathbf{k})\right].
\end{aligned}
$$

For example:

$$
\begin{aligned}
q &= [7,\ 2\mathbf{i} + 3\mathbf{j} + 4\mathbf{k}] \\
q^2 &= \left[7^2 - 2^2 - 3^2 - 4^2,\ 14(2\mathbf{i} + 3\mathbf{j} + 4\mathbf{k})\right] \\
&= [20,\ 28\mathbf{i} + 42\mathbf{j} + 56\mathbf{k}].
\end{aligned}
$$

The square of a pure quaternion is

$$
\begin{aligned}
\mathbf{v} &= x\mathbf{i} + y\mathbf{j} + z\mathbf{k} \\
q &= [0,\ \mathbf{v}] \\
q^2 &= [0,\ \mathbf{v}][0,\ \mathbf{v}] \\
&= [0 - \mathbf{v}\cdot\mathbf{v},\ \mathbf{v}\times\mathbf{v}] \\
&= [0 - \mathbf{v}\cdot\mathbf{v},\ \mathbf{0}] \\
&= \left[-\left(x^2 + y^2 + z^2\right),\ \mathbf{0}\right]
\end{aligned}
$$

which makes the square of a pure, unit-norm quaternion equal to -1, and was one of the results, to which some 19th-century mathematicians objected.

6.15.4 *Norm of the Quaternion Product*

In proving that the product of two unit-norm quaternions is another unit-norm quaternion we saw that

$$
\begin{aligned}
q_a &= [s_a,\ \mathbf{a}] \\
q_b &= [s_b,\ \mathbf{b}] \\
q_c &= q_a q_b \\
|q_c|^2 &= s_a^2\left(s_b^2 + b^2\right) + a^2\left(s_b^2 + b^2\right) \\
&= \left(s_a^2 + a^2\right)\left(s_b^2 + b^2\right)
\end{aligned}
$$

which, if we ignore the constraint of unit-norm quaternions, shows that the norm of a quaternion product equals the product of the individual norms:

$$
\begin{aligned}
|q_a q_b|^2 &= |q_a|^2 |q_b|^2 \\
|q_a q_b| &= |q_a||q_b|.
\end{aligned}
$$

6.16 Inverse Quaternion

An important feature of quaternion algebra is the ability to divide two quaternions q_b/q_a, as long as q_a does not vanish.

By definition, the inverse q^{-1} of q satisfies

$$qq^{-1} = [1,\ \mathbf{0}] = 1. \tag{6.12}$$

To isolate q^{-1}, we multiply (6.12) by q^*

$$
\begin{aligned}
q^* q q^{-1} &= q^* \\
|q|^2 q^{-1} &= q^*
\end{aligned} \tag{6.13}
$$

and from (6.13) we can write

$$q^{-1} = \frac{q^*}{|q|^2}.$$

If q is a unit-norm quaternion, then

$$q^{-1} = q^*$$

which is useful in the context of rotations.

Furthermore, as

$$(q_a q_b)^* = q_b^* q_a^*$$

then

$$(q_a q_b)^{-1} = q_b^{-1} q_a^{-1}.$$

Note that $qq^{-1} = q^{-1}q$:

$$
\begin{aligned}
qq^{-1} &= \frac{qq^*}{|q|^2} = 1 \\
q^{-1}q &= \frac{q^*q}{|q|^2} = 1.
\end{aligned}
$$

Thus, we represent the quotient q_b/q_a as

$$\begin{aligned} q_c &= \frac{q_b}{q_a} \\ &= q_b q_a^{-1} \\ &= \frac{q_b q_a^*}{|q_a|^2}. \end{aligned}$$

For completeness let's evaluate the inverse of q where

$$\begin{aligned} q &= \left[1,\ \tfrac{1}{\sqrt{3}}\mathbf{i} + \tfrac{1}{\sqrt{3}}\mathbf{j} + \tfrac{1}{\sqrt{3}}\mathbf{k}\right] \\ q^* &= \left[1,\ -\tfrac{1}{\sqrt{3}}\mathbf{i} - \tfrac{1}{\sqrt{3}}\mathbf{j} - \tfrac{1}{\sqrt{3}}\mathbf{k}\right] \\ |q|^2 &= 1 + \tfrac{1}{3} + \tfrac{1}{3} + \tfrac{1}{3} = 2 \\ q^{-1} &= \frac{q^*}{|q|^2} = \tfrac{1}{2}\left[1,\ -\tfrac{1}{\sqrt{3}}\mathbf{i} - \tfrac{1}{\sqrt{3}}\mathbf{j} - \tfrac{1}{\sqrt{3}}\mathbf{k}\right]. \end{aligned}$$

It should be clear that $q^{-1}q = 1$:

$$\begin{aligned} q^{-1}q &= \tfrac{1}{2}\left[1,\ -\tfrac{1}{\sqrt{3}}\mathbf{i} - \tfrac{1}{\sqrt{3}}\mathbf{j} - \tfrac{1}{\sqrt{3}}\mathbf{k}\right]\left[1,\ \tfrac{1}{\sqrt{3}}\mathbf{i} + \tfrac{1}{\sqrt{3}}\mathbf{j} + \tfrac{1}{\sqrt{3}}\mathbf{k}\right] \\ &= \tfrac{1}{2}\left[1 + \tfrac{1}{3} + \tfrac{1}{3} + \tfrac{1}{3},\ \mathbf{0}\right] \\ &= 1. \end{aligned}$$

6.17 Matrices

Matrices provide another way to express a quaternion product. For convenience, let's repeat (6.8) again and show it in matrix form:

$$\begin{aligned} [s_a,\ \mathbf{a}][s_b,\ \mathbf{b}] &= [s_a s_b - x_a x_b - y_a y_b - z_a z_b, \\ &\quad + s_a(x_b\mathbf{i} + y_b\mathbf{j} + z_b\mathbf{k}) + s_b(x_a\mathbf{i} + y_a\mathbf{j} + z_a\mathbf{k}) \\ &\quad + (y_a z_b - y_b z_a)\mathbf{i} + (z_a x_b - z_b x_a)\mathbf{j} + (x_a y_b - x_b y_a)\mathbf{k}] \\ &= \begin{bmatrix} s_a & -x_a & -y_a & -z_a \\ x_a & s_a & -z_a & y_a \\ y_a & z_a & s_a & -x_a \\ z_a & -y_a & x_a & s_a \end{bmatrix} \begin{bmatrix} s_b \\ x_b \\ y_b \\ z_b \end{bmatrix}. \end{aligned} \tag{6.14}$$

Let's recompute the product $q_a q_b$ using the above matrix:

$$
\begin{aligned}
q_a &= [1,\ 2\mathbf{i}+3\mathbf{j}+4\mathbf{k}] \\
q_b &= [2,\ 3\mathbf{i}+4\mathbf{j}+5\mathbf{k}] \\
q_a q_b &= \begin{bmatrix} 1 & -2 & -3 & -4 \\ 2 & 1 & -4 & 3 \\ 3 & 4 & 1 & -2 \\ 4 & -3 & 2 & 1 \end{bmatrix} \begin{bmatrix} 2 \\ 3 \\ 4 \\ 5 \end{bmatrix} \\
&= \begin{bmatrix} -36 \\ 6 \\ 12 \\ 12 \end{bmatrix} \\
&= [-36,\ 6\mathbf{i}+12\mathbf{j}+12\mathbf{k}].
\end{aligned}
$$

6.17.1 Orthogonal Matrix

We can demonstrate that the unit-norm quaternion matrix is orthogonal by showing that the product with its transpose equals the identity matrix. As we are dealing with matrices, $\mathbf{Q}$ will represent the matrix for q:

$$
\begin{aligned}
q &= [s,\ x\mathbf{i}+y\mathbf{j}+z\mathbf{k}] \\
\text{where}\quad 1 &= s^2+x^2+y^2+z^2 \\
\mathbf{Q} &= \begin{bmatrix} s & -x & -y & -z \\ x & s & -z & y \\ y & z & s & -x \\ z & -y & x & s \end{bmatrix} \\
\mathbf{Q}^{\mathrm{T}} &= \begin{bmatrix} s & x & y & z \\ -x & s & z & -y \\ -y & -z & s & x \\ -z & y & -x & s \end{bmatrix} \\
\mathbf{Q}\mathbf{Q}^{\mathrm{T}} &= \begin{bmatrix} s & -x & -y & -z \\ x & s & -z & y \\ y & z & s & -x \\ z & -y & x & s \end{bmatrix} \begin{bmatrix} s & x & y & z \\ -x & s & z & -y \\ -y & -z & s & x \\ -z & y & -x & s \end{bmatrix} \\
&= \begin{bmatrix} 1 & 0 & 0 & 0 \\ 0 & 1 & 0 & 0 \\ 0 & 0 & 1 & 0 \\ 0 & 0 & 0 & 1 \end{bmatrix}
\end{aligned}
$$

For this to occur, $\mathbf{Q}^{\mathrm{T}} = \mathbf{Q}^{-1}$.

6.18 Quaternion Algebra

Ordered pairs provide a simple notation for representing quaternions, and allow us to represent the real unit 1 as $[1, \ \mathbf{0}]$, and the imaginaries $i, \ j, \ k$ as $[0, \ \mathbf{i}]$, $[0, \ \mathbf{j}]$, $[0, \ \mathbf{k}]$ respectively. A quaternion then becomes a linear combination of these elements with associated real coefficients. Under such conditions, the elements form the *basis* for an algebra over the field of reals.

Furthermore, because quaternion algebra supports division, and obeys the normal axioms of algebra, except that multiplication is non-commutative, it is called a *division algebra*. The German mathematician Ferdinand Georg Frobenius proved that only three such real associative division algebras exist: real numbers, complex numbers and quaternions [8].

The *Cayley numbers* $\mathbb{O}$, constitute a real division algebra, but the Cayley numbers are 8-dimensional and are not associative, i.e. $a(bc) \neq (ab)c$ for all $a, b, c \in \mathbb{O}$.

6.19 Summary

Quaternions are very similar to complex numbers, apart from the fact that they have three imaginary terms, rather than one. Consequently, they inherit some of the properties associated with complex numbers, such as norm, complex conjugate, unit norm and inverse. They can also be added, subtracted, multiplied and divided. However, unlike complex numbers, they anti-commute when multiplied.

6.19.1 Summary of Definitions

Quaternion

$$
\begin{aligned}
q_a &= [s_a, \ \mathbf{a}] = [s_a, \ x_a\mathbf{i} + y_a\mathbf{j} + z_a\mathbf{k}] \\
q_b &= [s_b, \ \mathbf{b}] = [s_b, \ x_b\mathbf{i} + y_b\mathbf{j} + z_b\mathbf{k}] \,.
\end{aligned}
$$

Adding and subtracting

$$
q_a \pm q_b = [s_a \pm s_b, \ \mathbf{a} \pm \mathbf{b}].
$$

Product

$$
\begin{aligned}
q_a q_b &= [s_a, \ \mathbf{a}][s_b, \ \mathbf{b}] \\
&= [s_a s_b - \mathbf{a} \cdot \mathbf{b}, \ s_a\mathbf{b} + s_b\mathbf{a} + \mathbf{a} \times \mathbf{b}] \\
&= \begin{bmatrix} s_a & -x_a & -y_a & -z_a \\ x_a & s_a & -z_a & y_a \\ y_a & z_a & s_a & -x_a \\ z_a & -y_a & x_a & s_a \end{bmatrix} \begin{bmatrix} s_b \\ x_b \\ y_b \\ z_b \end{bmatrix}.
\end{aligned}
$$

Square

$$
\begin{aligned}
\mathbf{v} &= x\mathbf{i} + y\mathbf{j} + z\mathbf{k} \\
q^2 &= [s, \mathbf{v}][s, \mathbf{v}] \\
&= \left[s^2 - x^2 - y^2 - z^2, \ 2s(x\mathbf{i} + y\mathbf{j} + z\mathbf{k})\right].
\end{aligned}
$$

Pure

$$
\begin{aligned}
\mathbf{v} &= x\mathbf{i} + y\mathbf{j} + z\mathbf{k} \\
q^2 &= [0, \mathbf{v}][0, \mathbf{v}] \\
&= \left[-(x^2 + y^2 + z^2), \ \mathbf{0}\right].
\end{aligned}
$$

Norm

$$
\begin{aligned}
\mathbf{v} &= \lambda\hat{\mathbf{v}} \\
q &= [s, \lambda\hat{\mathbf{v}}] \\
|q| &= \sqrt{s^2 + \lambda^2}.
\end{aligned}
$$

Unit norm

$$
|q| = \sqrt{s^2 + \lambda^2} = 1.
$$

Conjugate

$$
\begin{aligned}
q^* &= [s, -\mathbf{v}] \\
(q_a q_b)^* &= q_b^* q_a^*.
\end{aligned}
$$

Inverse

$$
\begin{aligned}
q^{-1} &= \frac{q^*}{|q|^2} \\
(q_a q_b)^{-1} &= q_b^{-1} q_a^{-1}.
\end{aligned}
$$

6.20 Worked Examples

Here are some further worked examples that employ the ideas described above. In some cases, a test is included to confirm the result.

6.20.1 Adding and Subtracting Quaternions

Add and subtract the following quaternions:

$$\begin{aligned} q_a &= [2,\ -2\mathbf{i} + 3\mathbf{j} - 4\mathbf{k}] \\ q_b &= [1,\ -2\mathbf{i} + 5\mathbf{j} - 6\mathbf{k}] \\ q_a + q_b &= [3,\ -4\mathbf{i} + 8\mathbf{j} - 10\mathbf{k}] \\ q_a - q_b &= [1,\ 0\mathbf{i} - 2\mathbf{j} + 2\mathbf{k}]. \end{aligned}$$

6.20.2 Norm of a Quaternion

Find the norm of the following quaternions:

$$\begin{aligned} q_a &= [2,\ -2\mathbf{i} + 3\mathbf{j} - 4\mathbf{k}] \\ q_b &= [1,\ -2\mathbf{i} + 5\mathbf{j} - 6\mathbf{k}] \\ |q_a| &= \sqrt{2^2 + (-2)^2 + 3^2 + (-4)^2} = \sqrt{33} \\ |q_b| &= \sqrt{1^2 + (-2)^2 + 5^2 + (-6)^2} = \sqrt{66}. \end{aligned}$$

6.20.3 Unit-Norm Quaternions

Convert these quaternions to their unit-norm form:

$$\begin{aligned} q_a &= [2,\ -2\mathbf{i} + 3\mathbf{j} - 4\mathbf{k}] \\ q_b &= [1,\ -2\mathbf{i} + 5\mathbf{j} - 6\mathbf{k}] \\ |q_a| &= \sqrt{33} \\ |q_b| &= \sqrt{66} \\ q'_a &= \tfrac{1}{\sqrt{33}}[2,\ -2\mathbf{i} + 3\mathbf{j} - 4\mathbf{k}] \\ q'_b &= \tfrac{1}{\sqrt{66}}[1,\ -2\mathbf{i} + 5\mathbf{j} - 6\mathbf{k}]. \end{aligned}$$

6.20.4 Quaternion Product

Compute the product and reverse product of the following quaternions.

$$
\begin{aligned}
q_a &= [2,\ -2\mathbf{i} + 3\mathbf{j} - 4\mathbf{k}] \\
q_b &= [1,\ -2\mathbf{i} + 5\mathbf{j} - 6\mathbf{k}] \\
q_a q_b &= [2,\ -2\mathbf{i} + 3\mathbf{j} - 4\mathbf{k}][1,\ -2\mathbf{i} + 5\mathbf{j} - 6\mathbf{k}] \\
&= [2 \times 1 - ((-2) \times (-2) + 3 \times 5 + (-4) \times (-6)), \\
&\quad + 2(-2\mathbf{i} + 5\mathbf{j} - 6\mathbf{k}) + 1(-2\mathbf{i} + 3\mathbf{j} - 4\mathbf{k}) \\
&\quad + (3 \times (-6) - (-4) \times 5)\mathbf{i} - ((-2) \times (-6) - (-4) \times (-2))\mathbf{j} \\
&\quad + ((-2) \times 5 - 3 \times (-2))\mathbf{k}] \\
&= [-41,\ -6\mathbf{i} + 13\mathbf{j} - 16\mathbf{k} + 2\mathbf{i} - 4\mathbf{j} - 4\mathbf{k}] \\
&= [-41,\ -4\mathbf{i} + 9\mathbf{j} - 20\mathbf{k}].
\end{aligned}
$$

$$
\begin{aligned}
q_b q_a &= [1,\ -2\mathbf{i} + 5\mathbf{j} - 6\mathbf{k}][2 - 2\mathbf{i} + 3\mathbf{j} - 4\mathbf{k}] \\
&= [1 \times 2 - ((-2) \times (-2) + 5 \times 3 + (-6) \times (-4)), \\
&\quad + 1(-2\mathbf{i} + 3\mathbf{j} - 4\mathbf{k}) + 2(-2\mathbf{i} + 5\mathbf{j} - 6\mathbf{k}) \\
&\quad + (5 \times (-4) - (-6) \times 3)\mathbf{i} - ((-2) \times (-4) - (-6) \times (-2))\mathbf{j} \\
&\quad + ((-2) \times 3 - 5 \times (-2))\mathbf{k}] \\
&= [-41,\ -6\mathbf{i} + 13\mathbf{j} - 16\mathbf{k} - 2\mathbf{i} + 4\mathbf{j} + 4\mathbf{k}] \\
&= [-41,\ -8\mathbf{i} + 17\mathbf{j} - 12\mathbf{k}].
\end{aligned}
$$

Note: The only thing that has changed in this computation is the sign of the cross-product axial vector.

6.20.5 Square of a Quaternion

Compute the square of this quaternion:

$$
\begin{aligned}
q &= [2,\ -2\mathbf{i} + 3\mathbf{j} - 4\mathbf{k}] \\
q^2 &= [2,\ -2\mathbf{i} + 3\mathbf{j} - 4\mathbf{k}][2,\ -2\mathbf{i} + 3\mathbf{j} - 4\mathbf{k}] \\
&= [2 \times 2 - ((-2) \times (-2) + 3 \times 3 + (-4) \times (-4)), \\
&\quad + 2 \times 2(-2\mathbf{i} + 3\mathbf{j} - 4\mathbf{k})] \\
&= [-25,\ -8\mathbf{i} + 12\mathbf{j} - 16\mathbf{k}].
\end{aligned}
$$

6.20.6 *Inverse of a Quaternion*

Compute the inverse of this quaternion:

$$\begin{aligned} q &= [2,\ -2\mathbf{i} + 3\mathbf{j} - 4\mathbf{k}] \\ q^* &= [2,\ 2\mathbf{i} - 3\mathbf{j} + 4\mathbf{k}] \\ |q|^2 &= 2^2 + (-2)^2 + 3^2 + (-4)^2 = 33 \\ q^{-1} &= \tfrac{1}{33}[2,\ 2\mathbf{i} - 3\mathbf{j} + 4\mathbf{k}]. \end{aligned}$$

References

1. Hamilton, W.R.: On quaternions: or a new system of imaginaries in algebra. Phil. Mag. **3rd ser. 25** (1844)
2. Hamilton, W.R.: Lectures on Quaternions. Hodges and Smith, Dublin (1853)
3. Hamilton, W.R.: Elements of Quaternions (Jolly, C.J. (ed.) 2 vols.), 2nd edn. Green & Co., London, Longmans (1899–1901)
4. Tait, P.G.: An Elementary Treatise on Quaternions. Cambridge University Press, Cambridge (1867)
5. Gauss, C.F.: Mutation des Raumes In: Carl Friedrich Gauss Werke, Achter Band, pp. 357–361, König. Gesell. Wissen. Göttingen, 1900 (1819)
6. Wilson, E.B.: Vector Analysis. Yale University Press, New Haven (1901)
7. Feynman, R.P.: Symmetry and physical laws. In: Feynman Lectures in Physics, vol. 1
8. Altmann, S.L.: Rotations, : Quaternions and Double Groups, p. 16. Dover, New York (2005). ISBN-13: 978-0-486-44518–2

Chapter 7
3-D Rotation Transforms

7.1 Introduction

This chapter reviews the 3-D Euler rotation transforms employed in computer graphics software. In particular, we identify their *Achilles' heel*—gimbal lock—and the need to be able to rotate about an arbitrary axis. To this end, we will develop a matrix transform that achieves such a rotation, and in the following chapter develop a similar transform using quaternions.

7.2 3-D Rotation Transforms

The traditional technique for rotating points and frames of reference is based upon *Euler rotations*, named after the Swiss mathematician Leonhard Euler. They offer three ways to effect a rotation. The first is to rotate about one of the three Cartesian axes. The second combines any two of these rotations about different axes, and the third combines any three rotations.

Initially this approach sounds appealing—and in many cases works well—however, there are problems associated with the technique. The first problem is that when two or more rotations are combined, it is difficult to visualise and predict how the final rotation behaves. The second is that it is complicated to effect a rotation about a specific axis, and the third, is that under some conditions, one loses access to one of the object's rotational axes. This last problem is known as *gimbal lock*. To appreciate these issues we will construct a 3-D rotation transform that suffers from gimbal lock.

J. Vince, *Quaternions for Computer Graphics*,
https://doi.org/10.1007/978-1-4471-7509-4_7

7.3 Rotating About a Cartesian Axis

Figure 7.1 shows the geometry for rotating the point P, with the following:

- $P(x,\ y)$ is the point to be rotated about the origin,
- β is the angle to be rotated,
- θ is the angle P makes with the x-axis,
- $P'(x',\ y')$ is the rotated point.

To determine a general formula, we have from Fig. 7.1:

$$\begin{aligned}
\cos\theta &= \frac{x}{R}\\
\sin\theta &= \frac{y}{R}\\
x' &= R\cos(\theta+\beta)\\
&= R(\cos\theta\cos\beta - \sin\theta\sin\beta)\\
&= R\left(\frac{x}{R}\cos\beta - \frac{y}{R}\sin\beta\right)\\
&= x\cos\beta - y\sin\beta \qquad (7.1)\\
y' &= R\sin(\theta+\beta)\\
&= R(\sin\theta\cos\beta + \cos\theta\sin\beta)\\
&= R\left(\frac{y}{R}\cos\beta + \frac{x}{R}\sin\beta\right)\\
&= x\sin\beta + y\cos\beta. \qquad (7.2)
\end{aligned}$$

Equations (7.1) and (7.2) are represented by the matrix transform:

$$\begin{bmatrix} x' & y' \end{bmatrix} = \begin{bmatrix} \cos\beta & -\sin\beta \\ \sin\beta & \cos\beta \end{bmatrix} \begin{bmatrix} x \\ y \end{bmatrix}$$

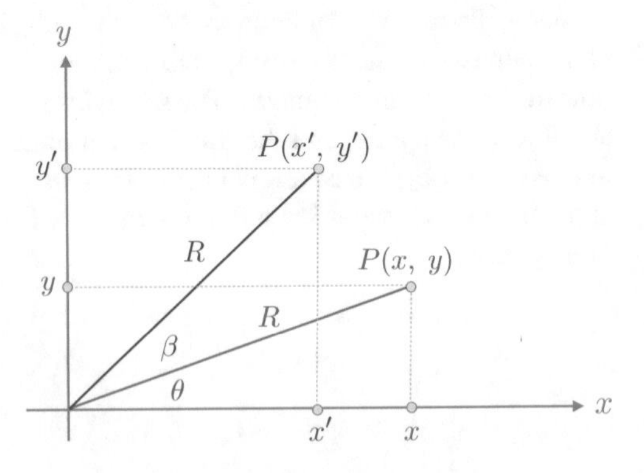

Fig. 7.1 The geometry associated with rotating a point about the origin

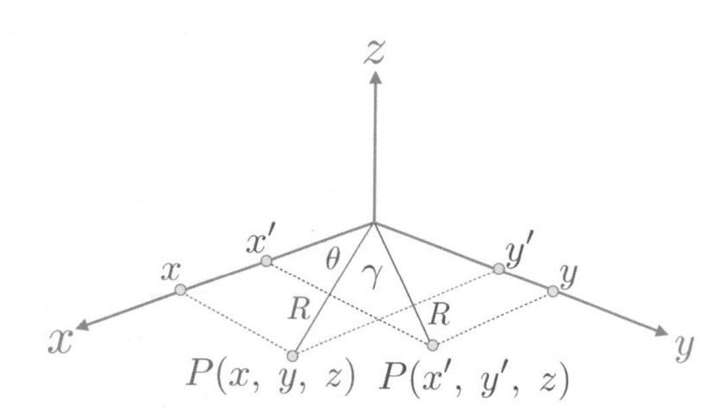

Fig. 7.2 Rotating a point about the z-axis

which is generalised into a 3-D rotation about a Cartesian axis by adding a third coordinate. For example, Fig. 7.2 shows the geometry to rotate the point $P(x, y, z)$ through an angle γ about the z-axis, where the z-coordinate remains untouched as follows:

$$\begin{bmatrix} x' & y' & z \end{bmatrix} = \begin{bmatrix} \cos\gamma & -\sin\gamma & 0 \\ \sin\gamma & \cos\gamma & 0 \\ 0 & 0 & 1 \end{bmatrix} \begin{bmatrix} x \\ y \\ z \end{bmatrix}.$$

For example, to rotate the point $(1, 1, 1)$ about the z-axis by 45° we use:

$$\begin{aligned} \begin{bmatrix} x' & y' & z \end{bmatrix} &= \begin{bmatrix} \cos 45^\circ & -\sin 45^\circ & 0 \\ \sin 45^\circ & \cos 45^\circ & 0 \\ 0 & 0 & 1 \end{bmatrix} \begin{bmatrix} 1 \\ 1 \\ 1 \end{bmatrix} \\ &\approx \begin{bmatrix} 0.707 & -0.707 & 0 \\ 0.707 & 0.707 & 0 \\ 0 & 0 & 1 \end{bmatrix} \begin{bmatrix} 1 \\ 1 \\ 1 \end{bmatrix} \\ &\approx [0 \quad 1.414 \quad 1]. \end{aligned}$$

To rotate a point about the x-axis, the x-coordinate remains constant whilst the y- and z-coordinates are changed according to (7.1) and (7.2):

$$P\left(x, y', z'\right) = \begin{bmatrix} 1 & 0 & 0 \\ 0 & \cos\alpha & -\sin\alpha \\ 0 & \sin\alpha & \cos\alpha \end{bmatrix} \begin{bmatrix} x \\ y \\ z \end{bmatrix}.$$

Finally, to rotate about the y-axis, the y-coordinate remains constant whilst the x- and z-coordinates are changed:

$$P(x', y, z') = \begin{bmatrix} \cos\beta & 0 & \sin\beta \\ 0 & 1 & 0 \\ -\sin\beta & 0 & \cos\beta \end{bmatrix} \begin{bmatrix} x \\ y \\ z \end{bmatrix}.$$

To summarise, we have three transforms to rotate a point about each Cartesian axis:

$$\mathbf{R}_{\alpha,\,x} = \begin{bmatrix} 1 & 0 & 0 \\ 0 & \cos\alpha & -\sin\alpha \\ 0 & \sin\alpha & \cos\alpha \end{bmatrix}$$

$$\mathbf{R}_{\beta,\,y} = \begin{bmatrix} \cos\beta & 0 & \sin\beta \\ 0 & 1 & 0 \\ -\sin\beta & 0 & \cos\beta \end{bmatrix}$$

$$\mathbf{R}_{\gamma,\,z} = \begin{bmatrix} \cos\gamma & -\sin\gamma & 0 \\ \sin\gamma & \cos\gamma & 0 \\ 0 & 0 & 1 \end{bmatrix}.$$

7.4 Rotate About an Off-Set Axis

To rotate about an axis parallel with one of the Cartesian axes, it is normal to employ homogeneous coordinates and translate the point to be rotated, such that it can be rotated about the origin, and then translated back by an equal and opposite amount. It is assumed that the reader is familiar with this strategy, but briefly, homogeneous coordinates permit addition to be incorporated within a matrix operation. For completeness, we will construct a transform that rotates a point about an axis parallel with the z−axis and intersects the point $(t_x,\ t_y,\ 0)$ as shown in Fig. 7.3:

$$\begin{bmatrix} x' \\ y' \\ z' \\ 1 \end{bmatrix} = \mathbf{T}_{(t_x,\,t_y,\,0)}\,\mathbf{R}_{\gamma,\,z}\,\mathbf{T}_{(-t_x,\,-t_y,\,0)} \begin{bmatrix} x \\ y \\ z \\ 1 \end{bmatrix}$$

where:

- $P(x,\ y,\ z)$ is the point to be rotated about an axis parallel with the z-axis,
- γ is the angle to be rotated,
- $(t_x,\ t_y,\ 0)$ locates the axis of rotation,
- $P'(x',\ y',\ z')$ is the rotated point,
- $\mathbf{T}_{(-t_x,\,-t_y,\,0)}$ is a translation matrix creating a temporary origin,
- $\mathbf{R}_{\gamma,\,z}$ is a rotation matrix,
- $\mathbf{T}_{(t_x,\,t_y,\,0)}$ is a translation matrix returning to the original origin.

The complete matrix transform is:

$$\mathbf{R}_{\gamma,\,z,\,(t_x,\,t_y,\,0)} = \begin{bmatrix} \cos\gamma & -\sin\gamma & 0 & t_x(1-\cos\gamma)+t_y\sin\gamma \\ \sin\gamma & \cos\gamma & 0 & t_y(1-\cos\gamma)-t_x\sin\gamma \\ 0 & 0 & 1 & 0 \\ 0 & 0 & 0 & 1 \end{bmatrix}.$$

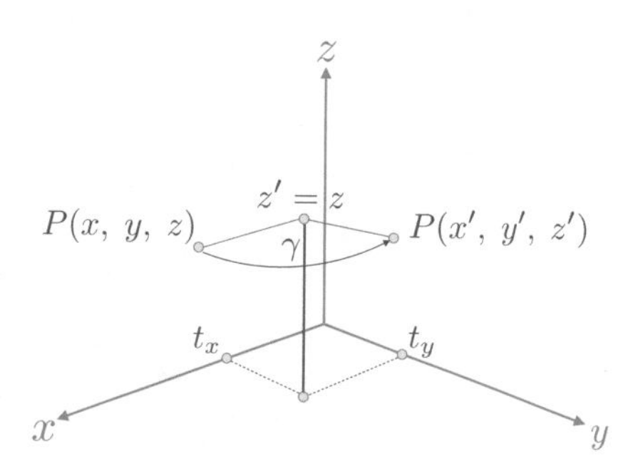

Fig. 7.3 Rotating a point about an axis parallel to the z-axis

The matrices for rotating about an off-set axis parallel with the x-axis and parallel with the y-axis are:

$$\mathbf{R}_{\alpha,\,x,\,(0,\,t_y,\,t_z)} = \begin{bmatrix} 1 & 0 & 0 & 0 \\ 0 & \cos\alpha & -\sin\alpha & t_y(1-\cos\alpha)+t_z\sin\alpha \\ 0 & \sin\alpha & \cos\alpha & t_z(1-\cos\alpha)-t_y\sin\alpha \\ 0 & 0 & 0 & 1 \end{bmatrix}.$$

$$\mathbf{R}_{\beta,\,y,\,(t_x,\,0,\,t_z)} = \begin{bmatrix} \cos\beta & 0 & \sin\beta & t_x(1-\cos\beta)-t_z\sin\beta \\ 0 & 1 & 0 & 0 \\ -\sin\beta & 0 & \cos\beta & t_z(1-\cos\beta)+t_x\sin\beta \\ 0 & 0 & 0 & 1 \end{bmatrix}.$$

7.5 Composite Rotations

Leaving aside the transforms for rotating about single and off-set axes, we have three transforms for rotating about the Cartesian axes: $\mathbf{R}_{\alpha,\,x}$, $\mathbf{R}_{\beta,\,y}$ and $\mathbf{R}_{\gamma,\,z}$, which can be combined to produce families of double and triple transforms. As mentioned above, such rotations are called *Euler rotations*, and it is assumed that the reader is familiar with their construction. The triple combinations are:

$$\begin{array}{llll} \mathbf{R}_{\gamma,\,x}\mathbf{R}_{\beta,\,y}\mathbf{R}_{\alpha,\,x} & \mathbf{R}_{\gamma,\,x}\mathbf{R}_{\beta,\,y}\mathbf{R}_{\alpha,\,z} & \mathbf{R}_{\gamma,\,x}\mathbf{R}_{\beta,\,z}\mathbf{R}_{\alpha,\,x} & \mathbf{R}_{\gamma,\,x}\mathbf{R}_{\beta,\,z}\mathbf{R}_{\alpha,\,y} \\ \mathbf{R}_{\gamma,\,y}\mathbf{R}_{\beta,\,x}\mathbf{R}_{\alpha,\,y} & \mathbf{R}_{\gamma,\,y}\mathbf{R}_{\beta,\,x}\mathbf{R}_{\alpha,\,z} & \mathbf{R}_{\gamma,\,y}\mathbf{R}_{\beta,\,z}\mathbf{R}_{\alpha,\,x} & \mathbf{R}_{\gamma,\,y}\mathbf{R}_{\beta,\,z}\mathbf{R}_{\alpha,\,y} \\ \mathbf{R}_{\gamma,\,z}\mathbf{R}_{\beta,\,x}\mathbf{R}_{\alpha,\,y} & \mathbf{R}_{\gamma,\,z}\mathbf{R}_{\beta,\,x}\mathbf{R}_{\alpha,\,z} & \mathbf{R}_{\gamma,\,z}\mathbf{R}_{\beta,\,y}\mathbf{R}_{\alpha,\,x} & \mathbf{R}_{\gamma,\,z}\mathbf{R}_{\beta,\,y}\mathbf{R}_{\alpha,\,z}. \end{array}$$

In order to illustrate the problem of gimbal lock we will employ a cube whose vertices are numbered 0 to 7 as shown in Fig. 7.4, and the coordinates in Table 7.1.

We can create a composite rotation transform by placing $\mathbf{R}_{\alpha,\,x}$, $\mathbf{R}_{\beta,\,y}$ and $\mathbf{R}_{\gamma,\,z}$ in any sequence—even repeating one of them twice, so long as they are separated by a different transform. As an example of the latter, we could use the sequence $\mathbf{R}_{\alpha,\,z}\mathbf{R}_{\beta,\,y}\mathbf{R}_{\gamma,\,z}$ where we rotate about the z-axis twice. However, to illustrate gimbal lock, let's choose the sequence $\mathbf{R}_{\gamma,\,z}\mathbf{R}_{\beta,\,y}\mathbf{R}_{\alpha,\,x}$ and make $\alpha = \beta = \gamma = 90°$, which

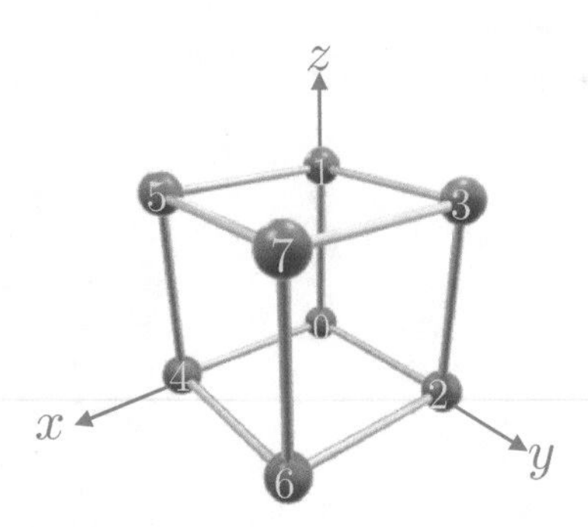

Fig. 7.4 A unit cube located at the origin

Table 7.1 Coordinates of the unit cube

Vertex	0	1	2	3	4	5	6	7
x	0	0	0	0	1	1	1	1
y	0	0	1	1	0	0	1	1
z	0	1	0	1	0	1	0	1

is equivalent to rotating a point 90° about the fixed x-axis, followed by a rotation of 90° about the fixed y-axis, followed by a rotation of 90° about the fixed z-axis. This rotation sequence is illustrated in Fig. 7.5. Note that as the rotation transforms are matrices, they are applied from right to left.

Figure 7.5a shows the starting position of the cube; b shows its position after a 90° rotation about the x-axis; c shows its position after a further rotation of 90° about the y-axis; and d the cube's resting position after a rotation of 90° about the z-axis.

However, in spite of employing three rotations about different axes, the cube has effectively only been rotated 90° about the y-axis! The cube has been rotated twice about the axis passing through vertices 0 and 4 and once about the axis passing through vertices 0 and 1, but the axis passing through vertices 0 and 2 has been ignored. This is known as *gimbal lock*, and arises through an unfortunate rotation sequence combination and angles. For those readers wanting a more detailed description of gimbal lock, please look at the author's book *Rotation Transforms for Computer Graphics* [1].

Reversing the composite rotation to $\mathbf{R}_{\alpha,x}\mathbf{R}_{\beta,y}\mathbf{R}_{\gamma,z}$ does not improve matters. This composite transform is equivalent to rotating a point 90° about the fixed z-axis, followed by a rotation of 90° about the fixed y-axis, followed by a rotation of 90° about the fixed x-axis. This rotation sequence is illustrated in Fig. 7.6.

Inspection of Fig. 7.6d shows that the unit cube has been rotated 180° about the vector $[1\ \ 0\ \ 1]^T$, i.e. an axis intersecting vertices 0 and 5. This time, the cube is rotated twice about an axis intersecting vertices 0 and 1, once about an axis intersecting vertices 0 and 4, and once again, the axis intersecting vertices 0 and 2 has been ignored. It is not difficult to see why Euler rotations cause so many problems. So let's continue and see how we can rotate about an arbitrary axis.

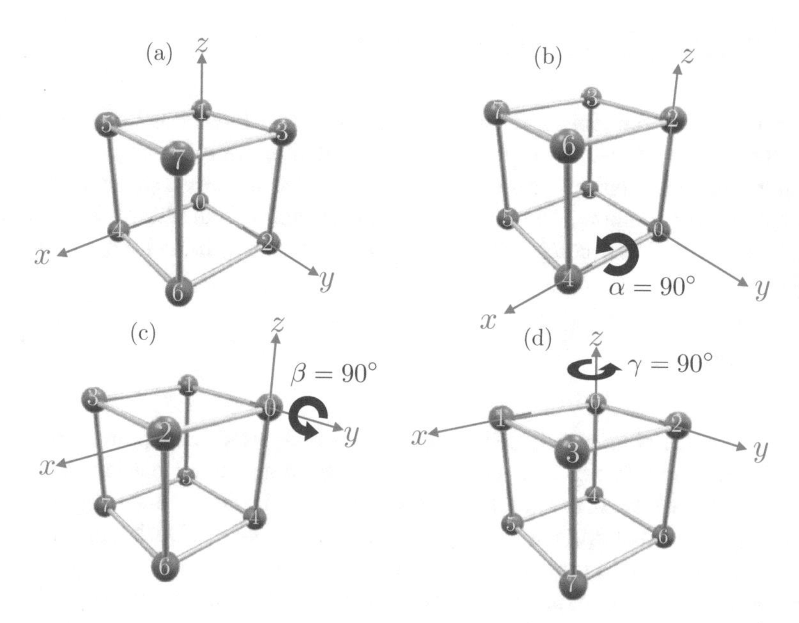

Fig. 7.5 Four views of the unit cube before and during the composite transform $\mathbf{R}_{\gamma,\ z}\mathbf{R}_{\beta,\ y}\mathbf{R}_{\alpha,\ x}$

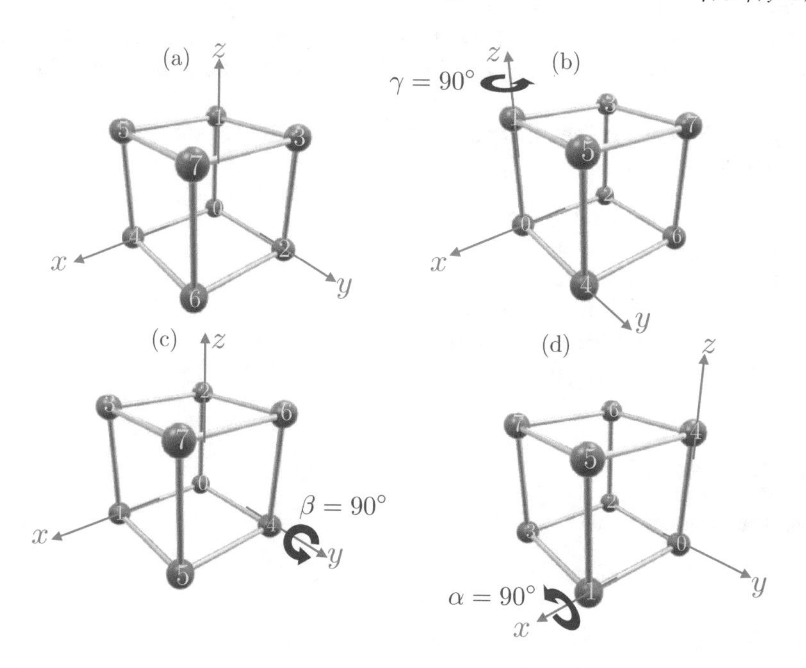

Fig. 7.6 Four views of the unit cube before and during the composite transform $\mathbf{R}_{\alpha,\ x}\mathbf{R}_{\beta,\ y}\mathbf{R}_{\gamma,\ z}$

7.6 Rotating About an Arbitrary Axis

There is nothing fundamentally wrong with individual Euler transforms—it is the way they are combined to effect a rotation that is flawed. Ideally, we require a rotation transform that permits us to specify the axis and angle of rotation, which is what we will compute. The first technique uses matrices and trigonometry and is rather laborious. The second approach employs vector analysis and is quite succinct.

7.6.1 *Matrices*

We begin by defining an axis using a unit vector $\hat{\mathbf{n}}$ about which a point P is rotated α to P' as shown in Fig. 7.7. And as we only have access to matrices that rotate points about the Cartesian axes, this unit vector has to be temporarily aligned with a Cartesian axis. In the following example we choose the x-axis. During the alignment process, the point P is subjected to the transforms necessary to align the unit vector with the x-axis. We then rotate P, α about the x-axis. To complete the operation, the rotated point is subjected to the transforms that return the unit vector to its original position.

Although matrices provide a powerful tool for undertaking this sort of work, it is, nevertheless, extremely tedious, but a good exercise for improving one's algebraic skills!

Figure 7.7 shows the geometry associated with rotating a point about an arbitrary axis, with the following:

- $P(x, y, z)$ is the point to be rotated about $\hat{\mathbf{n}}$,
- α is the angle to be rotated,
- $P'(x', y', z')$ is the rotated point,
- $\hat{\mathbf{n}} = a\mathbf{i} + b\mathbf{j} + c\mathbf{k}$ is the axis expressed as a unit vector,
- ϕ is the projected negative angle $\hat{\mathbf{n}}$ makes with the x-axis,
- θ is the angle $\hat{\mathbf{n}}$ makes with xz-plane.

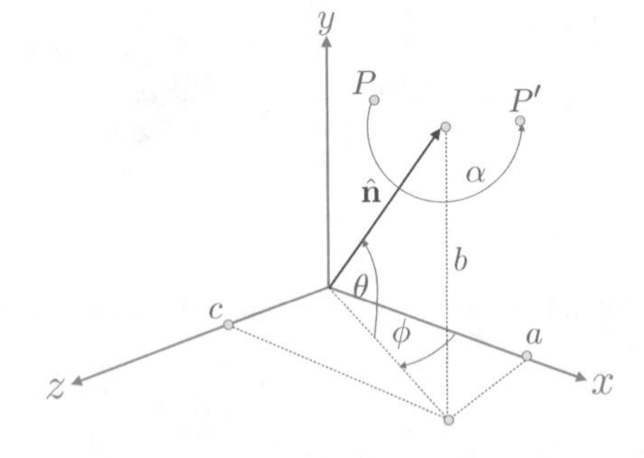

Fig. 7.7 The geometry associated with rotating a point about an arbitrary axis

The transforms to achieve this operation is expressed as follows:

$$\begin{bmatrix} x' \\ y' \\ z' \end{bmatrix} = \mathbf{R}_{-\phi,\, y}\, \mathbf{R}_{\theta,\, z}\, \mathbf{R}_{\alpha,\, x}\, \mathbf{R}_{-\theta,\, z}\, \mathbf{R}_{\phi,\, y} \begin{bmatrix} x \\ y \\ z \end{bmatrix}$$

which aligns the axis of rotation with the x-axis, performs the rotation of P through an angle α about the x-axis, and returns the axis of rotation back to its original position. Therefore,

$$\mathbf{R}_{\phi,\, y} = \begin{bmatrix} \cos\phi & 0 & \sin\phi \\ 0 & 1 & 0 \\ -\sin\phi & 0 & \cos\phi \end{bmatrix} \qquad \mathbf{R}_{-\theta,\, z} = \begin{bmatrix} \cos\theta & \sin\theta & 0 \\ -\sin\theta & \cos\theta & 0 \\ 0 & 0 & 1 \end{bmatrix}$$

$$\mathbf{R}_{\alpha,\, x} = \begin{bmatrix} 1 & 0 & 0 \\ 0 & \cos\alpha & -\sin\alpha \\ 0 & \sin\alpha & \cos\alpha \end{bmatrix} \qquad \mathbf{R}_{\theta,\, z} = \begin{bmatrix} \cos\theta & -\sin\theta & 0 \\ \sin\theta & \cos\theta & 0 \\ 0 & 0 & 1 \end{bmatrix}$$

$$\mathbf{R}_{-\phi,\, y} = \begin{bmatrix} \cos\phi & 0 & -\sin\phi \\ 0 & 1 & 0 \\ \sin\phi & 0 & \cos\phi \end{bmatrix}.$$

Let

$$\mathbf{R}_{-\phi,\, y}\, \mathbf{R}_{\theta,\, z}\, \mathbf{R}_{\alpha,\, x}\, \mathbf{R}_{-\theta,\, z}\, \mathbf{R}_{\phi,\, y} = \begin{bmatrix} a_{11} & a_{12} & a_{13} \\ a_{21} & a_{22} & a_{23} \\ a_{31} & a_{32} & a_{33} \end{bmatrix}$$

where by multiplying the matrices together we find that:

$$\begin{aligned}
a_{11} &= \cos^2\phi\cos^2\theta + \cos^2\phi\sin^2\theta\cos\alpha + \sin^2\phi\cos\alpha \\
a_{12} &= \cos\phi\cos\theta\sin\theta - \cos\phi\sin\theta\cos\theta\cos\alpha - \sin\phi\cos\theta\sin\alpha \\
a_{13} &= \cos\phi\sin\phi\cos^2\theta + \cos\phi\sin\phi\sin^2\theta\cos\alpha + \sin^2\phi\sin\theta\sin\alpha \\
&\quad + \cos^2\phi\sin\theta\sin\alpha - \cos\phi\sin\phi\cos\alpha \\
a_{21} &= \sin\theta\cos\theta\cos\phi - \cos\theta\sin\theta\cos\phi\cos\alpha + \cos\theta\sin\phi\sin\alpha \\
a_{22} &= \sin^2\theta + \cos^2\theta\cos\alpha \\
a_{23} &= \sin\theta\cos\theta\sin\phi - \cos\theta\sin\theta\sin\phi\cos\alpha - \cos\theta\cos\phi\sin\alpha \\
a_{31} &= \cos\phi\sin\phi\cos^2\theta + \cos\phi\sin\phi\sin^2\theta\cos\alpha - \cos^2\phi\sin\theta\sin\alpha \\
&= -\cos\phi\sin\phi\cos\alpha \\
a_{32} &= \sin\phi\cos\theta\sin\theta - \sin\phi\sin\theta\cos\theta\cos\alpha + \cos\phi\cos\theta\sin\alpha \\
a_{33} &= \sin^2\phi\cos^2\theta + \sin^2\phi\sin^2\theta\cos\alpha - \cos\phi\sin\phi\sin\theta\sin\alpha \\
&\quad + \cos\phi\sin\phi\sin\theta\sin\alpha + \cos^2\phi\cos\alpha.
\end{aligned}$$

After much trigonometric substitution we arrive at the transform

$$\begin{bmatrix} x'_p \\ y'_p \\ z'_p \end{bmatrix} = \begin{bmatrix} a^2K + \cos\alpha & abK - c\sin\alpha & acK + b\sin\alpha \\ abK + c\sin\alpha & b^2K + \cos\alpha & bcK - a\sin\alpha \\ acK - b\sin\alpha & bcK + a\sin\alpha & c^2K + \cos\alpha \end{bmatrix} \begin{bmatrix} x_p \\ y_p \\ z_p \end{bmatrix}$$

where,

$$K = 1 - \cos\alpha.$$

7.6.2 Vectors

Now let's solve the same problem using vectors. Figure 7.8 shows a view of the geometry associated with the task at hand. For clarification, Fig. 7.9 shows a cross-section and a plan view of the geometry. The following conditions apply for both figures:

- $P(x, y, z)$ is the point to be rotated about $\overrightarrow{ON}$,
- $\mathbf{p} = x\mathbf{i} + y\mathbf{j} + z\mathbf{k}$ is P's position vector,
- α is the angle to be rotated,
- $P'(x', y', z')$ is the rotated point,
- $\mathbf{p}' = x'\mathbf{i} + y'\mathbf{j} + z'\mathbf{k}$ is P''s position vector,
- $\hat{\mathbf{n}} = a\mathbf{i} + b\mathbf{j} + c\mathbf{k}$ is the axis expressed as a unit vector,
- θ is the angle $\mathbf{p}$ makes with $\hat{\mathbf{n}}$,
- $\mathbf{r}$ is a vector to any point on the circle of rotation,
- $\|\mathbf{r}\|$ is the distance of P and P' from the axis,
- O is the origin.

From Figs. 7.8 and 7.9:

$$\mathbf{p}' = \overrightarrow{ON} + \overrightarrow{NQ} + \overrightarrow{QP'}.$$

To find $\overrightarrow{ON}$:

$$\|\mathbf{n}\| = \|\mathbf{p}\| \cos\theta = \hat{\mathbf{n}} \cdot \mathbf{p}$$

therefore,

$$\overrightarrow{ON} = \mathbf{n} = \hat{\mathbf{n}}(\hat{\mathbf{n}} \cdot \mathbf{p}).$$

To find $\overrightarrow{NQ}$:

$$\overrightarrow{NQ} = \frac{NQ}{NP}\mathbf{r} = \frac{NQ}{NP'}\mathbf{r} = \cos\alpha\,\mathbf{r}$$

but

$$\mathbf{p} = \mathbf{n} + \mathbf{r} = \hat{\mathbf{n}}(\hat{\mathbf{n}} \cdot \mathbf{p}) + \mathbf{r}$$

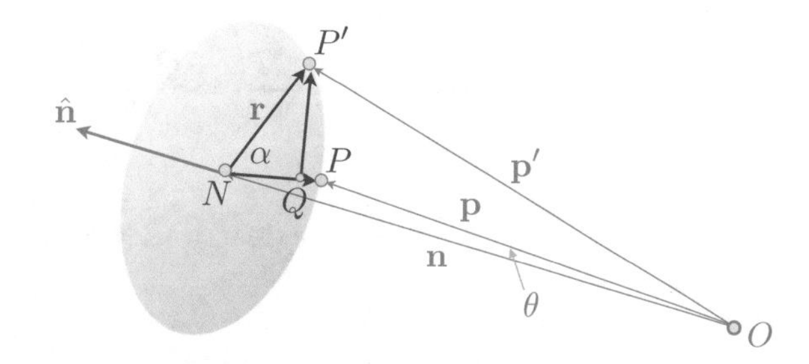

Fig. 7.8 The geometry associated with rotating a point about an arbitrary axis

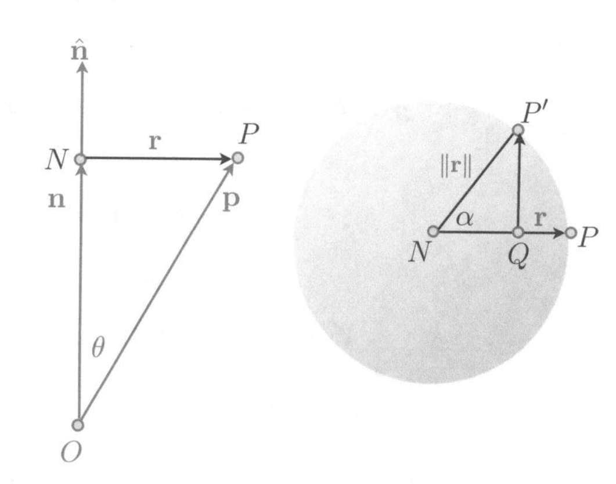

Fig. 7.9 A cross-section and plan view of the geometry associated with rotating a point about an arbitrary axis

therefore,

$$\mathbf{r} = \mathbf{p} - \hat{\mathbf{n}}(\hat{\mathbf{n}} \cdot \mathbf{p})$$

and

$$\overrightarrow{NQ} = [\mathbf{p} - \hat{\mathbf{n}}(\hat{\mathbf{n}} \cdot \mathbf{p})] \cos \alpha.$$

To find $\overrightarrow{QP'}$:
Let

$$\hat{\mathbf{n}} \times \mathbf{p} = \mathbf{w}$$

where,

$$\|\mathbf{w}\| = \|\hat{\mathbf{n}}\| \cdot \|\mathbf{p}\| \sin \theta = \|\mathbf{p}\| \sin \theta$$

but

$$\|\mathbf{r}\| = \|\mathbf{p}\| \sin \theta$$

therefore,

$$\|\mathbf{w}\| = \|\mathbf{r}\|.$$

Now

$$\frac{QP'}{NP'} = \frac{QP'}{\|\mathbf{r}\|} = \frac{QP'}{\|\mathbf{w}\|} = \sin\alpha$$

therefore,

$$\overrightarrow{QP'} = \mathbf{w}\sin\alpha = (\hat{\mathbf{n}} \times \mathbf{p})\sin\alpha$$

then,

$$\mathbf{p}' = \hat{\mathbf{n}}(\hat{\mathbf{n}} \cdot \mathbf{p}) + \big[\mathbf{p} - \hat{\mathbf{n}}(\hat{\mathbf{n}} \cdot \mathbf{p})\big]\cos\alpha + (\hat{\mathbf{n}} \times \mathbf{p})\sin\alpha$$

and

$$\mathbf{p}' = \mathbf{p}\cos\alpha + \hat{\mathbf{n}}(\hat{\mathbf{n}} \cdot \mathbf{p})(1 - \cos\alpha) + (\hat{\mathbf{n}} \times \mathbf{p})\sin\alpha.$$

Let

$$K = 1 - \cos\alpha$$

then,

$$\mathbf{p}' = \mathbf{p}\cos\alpha + \hat{\mathbf{n}}(\hat{\mathbf{n}} \cdot \mathbf{p})K + (\hat{\mathbf{n}} \times \mathbf{p})\sin\alpha. \tag{7.3}$$

Equation (7.3) expands to

$$\begin{aligned}
\mathbf{p}' &= (x_p\mathbf{i} + y_p\mathbf{j} + z_p\mathbf{k})\cos\alpha + (a\mathbf{i} + b\mathbf{j} + c\mathbf{k})(ax_p + by_p + cz_p)K \\
&\quad + \big[(bz_p - cy_p)\mathbf{i} + (cx_p - az_p)\mathbf{j} + (ay_p - bx_p)\mathbf{k}\big]\sin\alpha \\
\mathbf{p}' &= \big[x_p\cos\alpha + a(ax_p + by_p + cz_p)K + (bz_p - cy_p)\sin\alpha\big]\mathbf{i} \\
&\quad + \big[y_p\cos\alpha + b(ax_p + by_p + cz_p)K + (cx_p - az_p)\sin\alpha\big]\mathbf{j} \\
&\quad + \big[z_p\cos\alpha + c(ax_p + by_p + cz_p)K + (ay_p - bx_p)\sin\alpha\big]\mathbf{k} \\
\mathbf{p}' &= \big[x_p(a^2K + \cos\alpha) + y_p(abK - c\sin\alpha) + z_p(acK + b\sin\alpha)\big]\mathbf{i} \\
&\quad + \big[x_p(abK + c\sin\alpha) + y_p(b^2K + \cos\alpha) + z_p(bcK - a\sin\alpha)\big]\mathbf{j} \\
&\quad + \big[x_p(acK - b\sin\alpha) + y_p(bcK + a\sin\alpha) + z_p(c^2K + \cos\alpha)\big]\mathbf{k}
\end{aligned}$$

which unpacks into the transform:

$$\begin{bmatrix} x'_p \\ y'_p \\ z'_p \end{bmatrix} = \begin{bmatrix} a^2K + \cos\alpha & abK - c\sin\alpha & acK + b\sin\alpha \\ abK + c\sin\alpha & b^2K + \cos\alpha & bcK - a\sin\alpha \\ acK - b\sin\alpha & bcK + a\sin\alpha & c^2K + \cos\alpha \end{bmatrix} \begin{bmatrix} x_p \\ y_p \\ z_p \end{bmatrix}$$

where,

$$K = 1 - \cos\alpha$$

and is identical to the transform derived using matrices.

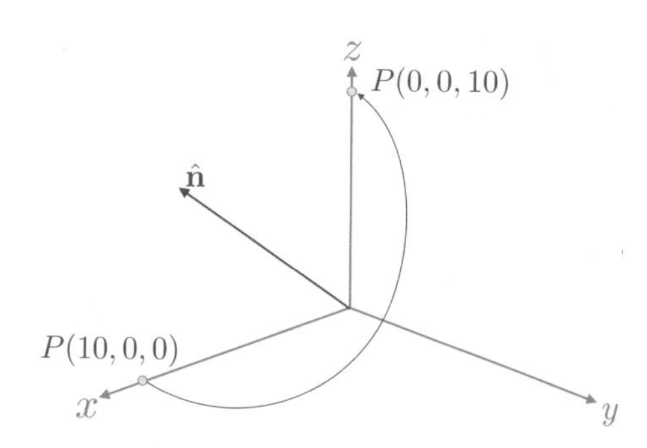

Fig. 7.10 Rotating the point P through 180° to P'

Let's test the transform with a simple example that can be easily verified. If we rotate the point $P(10,\ 0,\ 0)$, 180° about an axis defined by $\hat{\mathbf{n}} = \frac{1}{\sqrt{2}}\mathbf{i} + 0\mathbf{j} + \frac{1}{\sqrt{2}}\mathbf{k}$, it should be rotated to $P'(0,\ 0,\ 10)$ as shown in Fig. 7.10.

Therefore,

$$\alpha = 180^\circ, \quad \cos\alpha = -1, \quad \sin\alpha = 0, \quad K = 2$$

$$a = \frac{1}{\sqrt{2}}, \quad b = 0, \quad c = \frac{1}{\sqrt{2}}$$

and

$$\begin{bmatrix} 0 \\ 0 \\ 10 \end{bmatrix} = \begin{bmatrix} 0 & 0 & 1 \\ 0 & -1 & 0 \\ 1 & 0 & 0 \end{bmatrix} \begin{bmatrix} 10 \\ 0 \\ 0 \end{bmatrix}$$

which confirms our prediction.

7.7 Rodrigues' Rotation Formula

Benjamin Olinde Rodrigues was born in Bordeaux, France. He studied in Paris, and in 1816 was awarded his doctorate at the age of 21. The subject of his thesis was solving Legendre polynomials, and Rodrigues proposed a solution which is still known as the *Rodrigues formula*.

Although he pursued a career in politics and banking, his doctoral research confirms that he was more than just a 'recreational' mathematician, for in 1840 he published a mathematical paper in the *Annales de Mathématiques Pures et Appliquées* on transformation groups [2]. The paper contains a formula describing a geometric construction equating two successive rotations about different axes, with a third rotation about another axis. Today, we know this correspondence as the *Euler-Rodrigues Parameterisation*. Euler had already shown in 1775 that a single rotation could represent two successive rotations about different axes, but did not provide an algebraic solution [3].

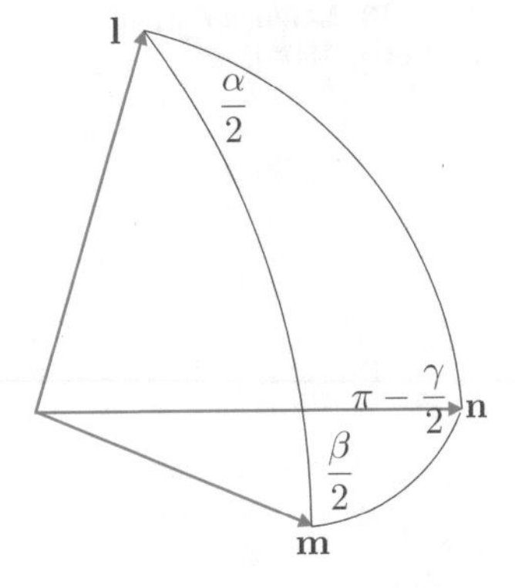

Fig. 7.11 Rodrigues' spherical triangle showing $R(\alpha l)$ and $R(\beta m)$ is the rotation, $R(\gamma n)$

If we represent a rotation α about an axial vector **a** as $\mathbf{R}_{\alpha,\,\mathbf{a}}$, then Rodrigues provided a solution to

$$\mathbf{R}_{\gamma,\,\mathbf{n}} = \mathbf{R}_{\alpha,\,\mathbf{l}}\mathbf{R}_{\beta,\,\mathbf{m}}$$

in the form of

$$\cos\left(\tfrac{\gamma}{2}\right) = \cos\left(\tfrac{\alpha}{2}\right)\cos\left(\tfrac{\beta}{2}\right) - \sin\left(\tfrac{\alpha}{2}\right)\sin\left(\tfrac{\beta}{2}\right)\mathbf{l}\cdot\mathbf{m} \tag{7.4}$$

$$\sin\left(\tfrac{\gamma}{2}\right)\mathbf{n} = \sin\left(\tfrac{\alpha}{2}\right)\cos\left(\tfrac{\beta}{2}\right)\mathbf{l} + \cos\left(\tfrac{\alpha}{2}\right)\sin\left(\tfrac{\beta}{2}\right)\mathbf{m} + \sin\left(\tfrac{\alpha}{2}\right)\sin\left(\tfrac{\beta}{2}\right)\mathbf{l}\times\mathbf{m}. \tag{7.5}$$

Rodrigues could not use the vector notation employed in (7.4) and (7.5), as this was yet to be defined by Gibbs and Hamilton [4], but he did employ the algebraic equivalent of these vector products, such as direction cosines. Figure 7.11 shows the spherical triangle formed by the axes and angles of rotation used by Rodrigues. Readers interested in learning more about this technique are guided towards the Internet [5].

Equations (7.4) and (7.5) contain some features familiar to the quaternion product, which become obvious with the following analysis. We start by defining the quaternions:

$$q_{\mathbf{l}} = \left[\cos\left(\tfrac{\alpha}{2}\right),\ \sin\left(\tfrac{\alpha}{2}\right)\mathbf{l}\right]$$
$$q_{\mathbf{m}} = \left[\cos\left(\tfrac{\beta}{2}\right),\ \sin\left(\tfrac{\beta}{2}\right)\mathbf{m}\right]$$
$$q_{\mathbf{n}} = \left[\cos\left(\tfrac{\gamma}{2}\right),\ \sin\left(\tfrac{\gamma}{2}\right)\mathbf{n}\right]$$

and form the product

$$\begin{aligned}
q_{\mathbf{n}} &= q_{\mathbf{l}} q_{\mathbf{m}} \\
&= \left[\cos\left(\tfrac{\alpha}{2}\right),\ \sin\left(\tfrac{\alpha}{2}\right)\mathbf{l}\right]\left[\cos\left(\frac{\beta}{2}\right),\ \sin\left(\frac{\beta}{2}\right)\mathbf{m}\right] \\
&= \Big[\cos\left(\tfrac{\alpha}{2}\right)\cos\left(\frac{\beta}{2}\right) - \sin\left(\tfrac{\alpha}{2}\right)\sin\left(\frac{\beta}{2}\right)\mathbf{l}\cdot\mathbf{m}, \\
&\quad \sin\left(\tfrac{\alpha}{2}\right)\cos\left(\frac{\beta}{2}\right)\mathbf{l} + \cos\left(\tfrac{\alpha}{2}\right)\sin\left(\frac{\beta}{2}\right)\mathbf{m} + \sin\left(\tfrac{\alpha}{2}\right)\sin\left(\frac{\beta}{2}\right)\mathbf{l}\times\mathbf{m}\Big]
\end{aligned}$$

$$\cos\left(\tfrac{\gamma}{2}\right) = \cos\left(\tfrac{\alpha}{2}\right)\cos\left(\frac{\beta}{2}\right) - \sin\left(\tfrac{\alpha}{2}\right)\sin\left(\frac{\beta}{2}\right)\mathbf{l}\cdot\mathbf{m} \tag{7.6}$$

$$\sin\left(\tfrac{\gamma}{2}\right)\mathbf{n} = \sin\left(\tfrac{\alpha}{2}\right)\cos\left(\frac{\beta}{2}\right)\mathbf{l} + \cos\left(\tfrac{\alpha}{2}\right)\sin\left(\frac{\beta}{2}\right)\mathbf{m} + \sin\left(\tfrac{\alpha}{2}\right)\sin\left(\frac{\beta}{2}\right)\mathbf{l}\times\mathbf{m} \tag{7.7}$$

where (7.6) and (7.7) are identical to (7.4) and (7.5) respectively. Although Rodrigues had not invented quaternions in the form of

$$q = s + ai + bj + ck,$$

he had discovered the coefficients of a quaternion product before Hamilton. *C'est la vie!*

Today, Rodrigues' rotation formula is expressed:

$$\mathbf{p}' = \mathbf{p}\cos\alpha + \hat{\mathbf{n}}(\hat{\mathbf{n}}\cdot\mathbf{p})(1-\cos\alpha) + (\hat{\mathbf{n}}\times\mathbf{p})\sin\alpha \tag{7.8}$$

where,

- $\hat{\mathbf{n}}$ is the unit vector defining the arbitrary axis,
- $\mathbf{p}$ is the vector to be rotated,
- α is the angle of rotation,
- $\mathbf{p}'$ is the rotated vector.

Let's illustrate (7.8) with an example where vector $\mathbf{p} = [10\quad 10\quad 10]^{\mathrm{T}}$ is rotated 90° about the z-axis. This makes $\hat{\mathbf{n}} = [0\quad 0\quad 1]^{\mathrm{T}}$ and $\alpha = 90°$. Therefore, $\cos\alpha = 0$ and $\sin\alpha = 1$:

$$\begin{aligned}
\mathbf{p}' &= \mathbf{p}\cos\alpha + \hat{\mathbf{n}}(\hat{\mathbf{n}}\cdot\mathbf{p})(1-\cos\alpha) + (\hat{\mathbf{n}}\times\mathbf{p})\sin\alpha \\
&= 0 + [0\quad 0\quad 1]([0\quad 0\quad 1]\cdot[10\quad 10\quad 10]) + [0\quad 0\quad 1]\times[10\quad 10\quad 10] \\
&= [0\quad 0\quad 1]10 + [-10\quad 10\quad 0] \\
&= [-10\quad 10\quad 10].
\end{aligned}$$

Which is correct.

Rodrigues' rotation formula can also be expressed as a rotation matrix:

$$\mathbf{R} = \mathbf{I} + (\sin\alpha)\mathbf{K} + (1-\cos\alpha)\mathbf{K}^2 \tag{7.9}$$

where,

- $\mathbf{R}$ is the rotation matrix,
- $\mathbf{I}$ is the identity matrix,
- α is the angle of rotation,
- $\hat{\mathbf{n}} = n_x\mathbf{i} + n_y\mathbf{j} + n_z\mathbf{k}$ is the unit vector defining the arbitrary axis,
- $\mathbf{K}$ is the cross-product matrix:

$$\mathbf{K} = \begin{bmatrix} 0 & -n_z & n_y \\ n_z & 0 & -n_x \\ -n_y & n_x & 0 \end{bmatrix}.$$

Let's illustrate (7.9) with the previous example where vector $\mathbf{p} = [10 \quad 10 \quad 10]$ is rotated 90° about the z-axis. This makes $\hat{\mathbf{n}} = [0 \quad 0 \quad 1]^{\mathrm{T}}$ and $\alpha = 90°$. Therefore, $\cos\alpha = 0$ and $\sin\alpha = 1$:

$$\begin{aligned}
\mathbf{K} &= \begin{bmatrix} 0 & -1 & 0 \\ 1 & 0 & 0 \\ 0 & 0 & 0 \end{bmatrix} \\
\mathbf{K}^2 &= \begin{bmatrix} 0 & -1 & 0 \\ 1 & 0 & 0 \\ 0 & 0 & 0 \end{bmatrix} \begin{bmatrix} 0 & -1 & 0 \\ 1 & 0 & 0 \\ 0 & 0 & 0 \end{bmatrix} = \begin{bmatrix} -1 & 0 & 0 \\ 0 & -1 & 0 \\ 0 & 0 & 0 \end{bmatrix} \\
\mathbf{R} &= \mathbf{I} + \mathbf{K} + \mathbf{K}^2 \\
&= \begin{bmatrix} 1 & 0 & 0 \\ 0 & 1 & 0 \\ 0 & 0 & 1 \end{bmatrix} + \begin{bmatrix} 0 & -1 & 0 \\ 1 & 0 & 0 \\ 0 & 0 & 0 \end{bmatrix} + \begin{bmatrix} -1 & 0 & 0 \\ 0 & -1 & 0 \\ 0 & 0 & 0 \end{bmatrix} \\
&= \begin{bmatrix} 0 & -1 & 0 \\ 1 & 0 & 0 \\ 0 & 0 & 1 \end{bmatrix}.
\end{aligned}$$

Next, we apply this matrix to rotate the point $\mathbf{p}$:

$$\begin{aligned}
\mathbf{p}' &= \begin{bmatrix} 0 & -1 & 0 \\ 1 & 0 & 0 \\ 0 & 0 & 1 \end{bmatrix} \begin{bmatrix} 10 \\ 10 \\ 10 \end{bmatrix} \\
&= \begin{bmatrix} -10 \\ 10 \\ 10 \end{bmatrix}.
\end{aligned}$$

Which agrees with the original example.

7.8 Summary

In this chapter we have reviewed the matrix transforms for rotating a point about one of the three Cartesian axes. By employing homogeneous coordinates, the translation transform can be integrated to rotate points about an off-set axis parallel with one of the Cartesian axes.

Composite rotations are created by combining the matrices representing the individual rotations about three successive axes. Such rotations are known as Euler rotations, and there are twelve ways of combining these matrices. Unfortunately, one of the problems with such transforms is that they suffer from gimbal lock, where one degree of freedom is lost under certain angle combinations. Another problem, is that it is difficult to predict how a point moves in space when animated by a composite transform, although they are widely used for positioning objects in world space.

Finally, matrices and vectors were used to develop a transform for rotating a point about an arbitrary axis.

7.8.1 Summary of Definitions

Translate a point

$$\mathbf{T}_{(t_x,\,t_y,\,t_z)} = \begin{bmatrix} 1 & 0 & 0 & t_x \\ 0 & 1 & 0 & t_y \\ 0 & 0 & 1 & t_z \\ 0 & 0 & 0 & 1 \end{bmatrix}.$$

Rotate a point about the $x-$, $y-$, $z-$ axis

$$\mathbf{R}_{\alpha,\,x} = \begin{bmatrix} 1 & 0 & 0 \\ 0 & \cos\alpha & -\sin\alpha \\ 0 & \sin\alpha & \cos\alpha \end{bmatrix}$$

$$\mathbf{R}_{\alpha,\,y} = \begin{bmatrix} \cos\alpha & 0 & \sin\alpha \\ 0 & 1 & 0 \\ -\sin\alpha & 0 & \cos\alpha \end{bmatrix}$$

$$\mathbf{R}_{\alpha,\,z} = \begin{bmatrix} \cos\alpha & -\sin\alpha & 0 \\ \sin\alpha & \cos\alpha & 0 \\ 0 & 0 & 1 \end{bmatrix}.$$

Rotate a point about off-set $x-$, $y-$, $z-$ axis

$$\mathbf{R}_{\alpha,\ x,\ (0,\ t_y,\ t_z)} = \begin{bmatrix} 1 & 0 & 0 & 0 \\ 0 & \cos\alpha & -\sin\alpha & t_y(1-\cos\alpha)+t_z\sin\alpha \\ 0 & \sin\alpha & \cos\alpha & t_z(1-\cos\alpha)-t_y\sin\alpha \\ 0 & 0 & 0 & 1 \end{bmatrix}$$

$$\mathbf{R}_{\alpha,\ y,\ (t_x,\ 0,\ t_z)} = \begin{bmatrix} \cos\alpha & 0 & \sin\alpha & t_x(1-\cos\alpha)-t_z\sin\alpha \\ 0 & 1 & 0 & 0 \\ -\sin\alpha & 0 & \cos\alpha & t_z(1-\cos\alpha)+t_x\sin\alpha \\ 0 & 0 & 0 & 1 \end{bmatrix}$$

$$\mathbf{R}_{\alpha,\ z,\ (t_x,\ t_y,\ 0)} = \begin{bmatrix} \cos\alpha & -\sin\alpha & 0 & t_x(1-\cos\alpha)+t_y\sin\alpha \\ \sin\alpha & \cos\alpha & 0 & t_y(1-\cos\alpha)-t_x\sin\alpha \\ 0 & 0 & 1 & 0 \\ 0 & 0 & 0 & 1 \end{bmatrix}.$$

Rotate a point about an arbitrary axis

$$\mathbf{R}_{\alpha,\ \hat{\mathbf{n}}} = \begin{bmatrix} a^2K+\cos\alpha & abK-c\sin\alpha & acK+b\sin\alpha \\ abK+c\sin\alpha & b^2K+\cos\alpha & bcK-a\sin\alpha \\ acK-b\sin\alpha & bcK+a\sin\alpha & c^2K+\cos\alpha \end{bmatrix}$$

$$K = 1-\cos\alpha$$

$$\hat{\mathbf{n}} = a\mathbf{i}+b\mathbf{j}+c\mathbf{k}.$$

7.9 Worked Examples

Here are some worked examples that employ the ideas described above.

7.9.1 Rotation Transform About an Off-Set Axis

Develop a rotation transform to rotate a point about an axis off-set to the y-axis.

Let the off-set axis intersect the point $(t_x,\ 0,\ t_z)$. Therefore, the homogeneous transform for this rotation is

$$\begin{bmatrix} x' \\ y' \\ z' \\ 1 \end{bmatrix} = \mathbf{T}_{(t_x,\ 0,\ t_z)}\,\mathbf{R}_{\alpha,\ y}\,\mathbf{T}_{(-t_x,\ 0,\ -t_z)} \begin{bmatrix} x \\ y \\ z \\ 1 \end{bmatrix}$$

where

$\mathbf{T}_{(-t_x,\,0,\,-t_z)}$ creates a temporary origin,
$\mathbf{R}_{\alpha,\,y}$ rotates α about the temporary y-axis,
$\mathbf{T}_{(t_x,\,0,\,t_z)}$ returns to the original position,

and

$$\mathbf{T}_{(t_x,\,0,\,t_z)} = \begin{bmatrix} 1 & 0 & 0 & t_x \\ 0 & 1 & 0 & 0 \\ 0 & 0 & 1 & t_z \\ 0 & 0 & 0 & 1 \end{bmatrix}$$

$$\mathbf{T}_{(-t_x,\,0,\,-t_z)} = \begin{bmatrix} 1 & 0 & 0 & -t_x \\ 0 & 1 & 0 & 0 \\ 0 & 0 & 1 & -t_z \\ 0 & 0 & 0 & 1 \end{bmatrix}$$

$$\mathbf{R}_{\alpha,\,y} = \begin{bmatrix} \cos\alpha & 0 & \sin\alpha & 0 \\ 0 & 1 & 0 & 0 \\ -\sin\alpha & 0 & \cos\alpha & 0 \\ 0 & 0 & 0 & 1 \end{bmatrix}.$$

Therefore,

$$\mathbf{T}_{(t_x,\,0,\,t_z)}\mathbf{R}_{\alpha,\,y}\mathbf{T}_{(-t_x,\,0,\,-t_z)} =$$

$$= \begin{bmatrix} 1 & 0 & 0 & t_x \\ 0 & 1 & 0 & 0 \\ 0 & 0 & 1 & t_z \\ 0 & 0 & 0 & 1 \end{bmatrix} \begin{bmatrix} \cos\alpha & 0 & \sin\alpha & 0 \\ 0 & 1 & 0 & 0 \\ -\sin\alpha & 0 & \cos\alpha & 0 \\ 0 & 0 & 0 & 1 \end{bmatrix} \begin{bmatrix} 1 & 0 & 0 & -t_x \\ 0 & 1 & 0 & 0 \\ 0 & 0 & 1 & -t_z \\ 0 & 0 & 0 & 1 \end{bmatrix}$$

$$= \begin{bmatrix} 1 & 0 & 0 & t_x \\ 0 & 1 & 0 & 0 \\ 0 & 0 & 1 & t_z \\ 0 & 0 & 0 & 1 \end{bmatrix} \begin{bmatrix} \cos\alpha & 0 & \sin\alpha & -t_x\cos\alpha - t_z\sin\alpha \\ 0 & 1 & 0 & 0 \\ -\sin\alpha & 0 & \cos\alpha & t_x\sin\alpha - t_z\cos\alpha \\ 0 & 0 & 0 & 1 \end{bmatrix}$$

$$= \begin{bmatrix} \cos\alpha & 0 & \sin\alpha & t_x(1-\cos\alpha) - t_z\sin\alpha \\ 0 & 1 & 0 & 0 \\ -\sin\alpha & 0 & \cos\alpha & t_z(1-\cos\alpha) + t_x\sin\alpha \\ 0 & 0 & 0 & 1 \end{bmatrix}.$$

7.9.2 *Test for Gimbal Lock*

Compute the rotation transform $\mathbf{R}_{\gamma,\,x}\mathbf{R}_{\beta,\,y}\mathbf{R}_{\alpha,\,x}$ and see if it suffers from gimbal lock when $\alpha = \beta = \gamma = 90°$. What is the axis and angle of rotation?
Using the notation $c_\beta = \cos\beta$ and $s_\beta = \sin\beta$, the composite transform is

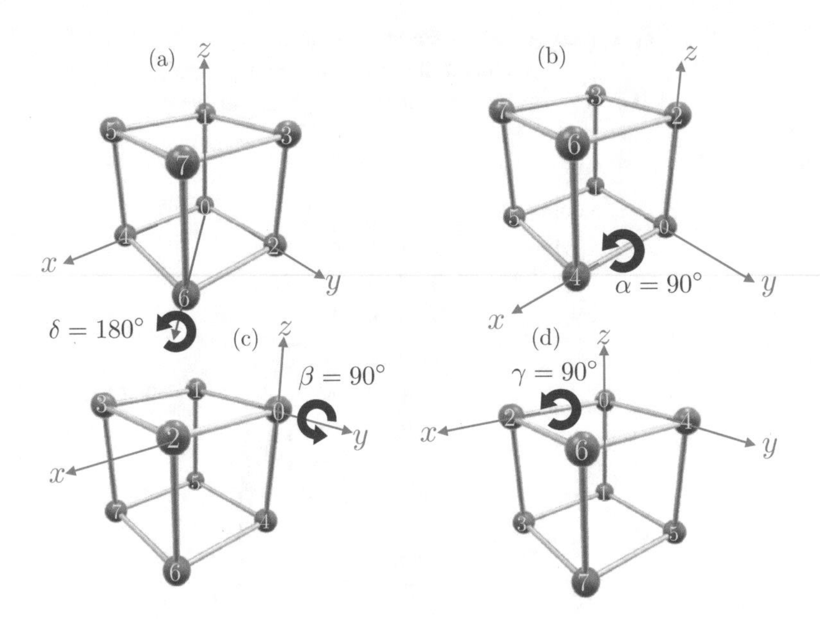

Fig. 7.12 Four views of the unit cube before and during the composite transform $\mathbf{R}_{\gamma,\,x}\mathbf{R}_{\beta,\,y}\mathbf{R}_{\alpha,\,x}$

$$
\begin{aligned}
\mathbf{R}_{\gamma,\,x}\mathbf{R}_{\beta,\,y}\mathbf{R}_{\alpha,\,x} &= \begin{bmatrix} 1 & 0 & 0 \\ 0 & c_\gamma & -s_\gamma \\ 0 & s_\gamma & c_\gamma \end{bmatrix} \begin{bmatrix} c_\beta & 0 & s_\beta \\ 0 & 1 & 0 \\ -s_\beta & 0 & c_\beta \end{bmatrix} \begin{bmatrix} 1 & 0 & 0 \\ 0 & c_\alpha & -s_\alpha \\ 0 & s_\alpha & c_\alpha \end{bmatrix} \\
&= \begin{bmatrix} c_\beta & s_\beta s_\alpha & s_\beta c_\alpha \\ s_\gamma s_\beta & (c_\gamma c_\alpha - s_\gamma c_\beta s_\alpha) & (-c_\gamma s_\alpha - s_\gamma c_\beta c_\alpha) \\ -c_\gamma s_\beta & (s_\gamma c_\alpha + c_\gamma c_\beta s_\alpha) & (-s_\gamma s_\alpha + c_\gamma c_\beta c_\alpha) \end{bmatrix} \\
\mathbf{R}_{90^\circ,\,x}\mathbf{R}_{90^\circ,\,y}\mathbf{R}_{90^\circ,\,x} &= \begin{bmatrix} 0 & 1 & 0 \\ 1 & 0 & 0 \\ 0 & 0 & -1 \end{bmatrix}.
\end{aligned}
$$

Figure 7.12 shows a cube at each stage of rotation, and it is clear that gimbal lock is not present as the cube is rotated through each of its orthogonal axes. The axis of rotation is through the vertices 0 and 6, i.e. $[1 \quad 1 \quad 0]^{\mathrm{T}}$ and the angle of rotation is 180°.

7.9.3 The General Rotation Matrix

Show that the rotation matrix for rotating points about an arbitrary axis works for the three Cartesian axes.

Starting with the matrix:

$$\mathbf{R}_{\alpha,\,\hat{\mathbf{n}}} = \begin{bmatrix} a^2K + \cos\alpha & abK - c\sin\alpha & acK + b\sin\alpha \\ abK + c\sin\alpha & b^2K + \cos\alpha & bcK - a\sin\alpha \\ acK - b\sin\alpha & bcK + a\sin\alpha & c^2K + \cos\alpha \end{bmatrix}$$

$$K = 1 - \cos\alpha$$

$$\hat{\mathbf{n}} = a\mathbf{i} + b\mathbf{j} + c\mathbf{k}.$$

Rotating about the x-axis:

$$\hat{\mathbf{n}} = a\mathbf{i}$$

therefore, $a = 1$ and $b = c = 0$:

$$\mathbf{R}_{\alpha,\,x} = \begin{bmatrix} 1 & 0 & 0 \\ 0 & \cos\alpha & -\sin\alpha \\ 0 & \sin\alpha & \cos\alpha \end{bmatrix}.$$

Rotating about the y-axis:

$$\hat{\mathbf{n}} = b\mathbf{j}$$

therefore, $b = 1$ and $a = c = 0$:

$$\mathbf{R}_{\alpha,\,y} = \begin{bmatrix} \cos\alpha & 0 & \sin\alpha \\ 0 & 1 & 0 \\ -\sin\alpha & 0 & \cos\alpha \end{bmatrix}.$$

Rotating about the z-axis:

$$\hat{\mathbf{n}} = c\mathbf{k}$$

therefore, $c = 1$ and $a = b = 0$:

$$\mathbf{R}_{\alpha,\,z} = \begin{bmatrix} \cos\alpha & -\sin\alpha & 0 \\ \sin\alpha & \cos\alpha & 0 \\ 0 & 0 & 1 \end{bmatrix}.$$

Which are correct.

7.9.4 Testing the Rotation Matrix

Compute the rotation transform to rotate a point 180° about an axis aligned with $[1\ \ 1\ \ 1]^T$. Show by example, that rotating a point twice by this transform returns it to its original position.

Starting with the matrix:

$$\mathbf{R}_{\alpha,\,\hat{\mathbf{n}}} = \begin{bmatrix} a^2K + \cos\alpha & abK - c\sin\alpha & acK + b\sin\alpha \\ abK + c\sin\alpha & b^2K + \cos\alpha & bcK - a\sin\alpha \\ acK - b\sin\alpha & bcK + a\sin\alpha & c^2K + \cos\alpha \end{bmatrix}$$

$$K = 1 - \cos\alpha$$

$$\hat{\mathbf{n}} = a\mathbf{i} + b\mathbf{j} + c\mathbf{k}.$$

Therefore, given

$$\mathbf{n} = 1\mathbf{i} + 1\mathbf{j} + 1\mathbf{k}$$

$$\hat{\mathbf{n}} = \frac{1}{\sqrt{3}}\mathbf{i} + \frac{1}{\sqrt{3}}\mathbf{j} + \frac{1}{\sqrt{3}}\mathbf{k}$$

and

$$a = b = c = \frac{1}{\sqrt{3}}.$$

Given $\alpha = 180°$, $\cos\alpha = -1$, $\sin\alpha = 0$ and $K = 2$, and the matrix becomes:

$$\mathbf{R}_{180°,\,\hat{\mathbf{n}}} = \begin{bmatrix} -1/3 & 2/3 & 2/3 \\ 2/3 & -1/3 & 2/3 \\ 2/3 & 2/3 & -1/3 \end{bmatrix}.$$

Multiplying this matrix by itself must result in the identity matrix:

$$\mathbf{R}_{180°,\,\hat{\mathbf{n}}}\mathbf{R}_{180°,\,\hat{\mathbf{n}}} = \begin{bmatrix} -1/3 & 2/3 & 2/3 \\ 2/3 & -1/3 & 2/3 \\ 2/3 & 2/3 & -1/3 \end{bmatrix}\begin{bmatrix} -1/3 & 2/3 & 2/3 \\ 2/3 & -1/3 & 2/3 \\ 2/3 & 2/3 & -1/3 \end{bmatrix}$$

$$= \begin{bmatrix} 1 & 0 & 0 \\ 0 & 1 & 0 \\ 0 & 0 & 1 \end{bmatrix}.$$

Which confirms that any point rotated twice by the rotation matrix returns to its original point.

References

1. Vince, J.A.: Rotation Transforms for Computer Graphics. Springer, Berlin (2011). ISBN 978-0-85729-153-0
2. Rodrigues, B.O.: Des lois géométriques qui régissent les déplacements d'un système solide dans l'espace, et de la variation des coordonnées provent de ses déplacements considérés indépendamment des causes qui peuvent les produire. J. de Matématiques Pures et Appliquées **5**, 380–440 (1840)

3. Eulero, L.: Nova methodus motum corporum rigidorum determinandi Novi Comment. Acad. Sci. Imp. Petropolitanae **20**, 20838 (1775)
4. Gibbs, J.W.: Vector Analysis. A Text-book for the Use of Students of Mathematics and Physics. Yale University Press (1901)
5. Visualizing Rotations and Composition of Rotations with Rodrigues' Vector. www.arxiv.org

Chapter 8
Quaternions in Space

8.1 Introduction

In this chapter we show how quaternions are used to rotate vectors about an arbitrary axis. We begin by reviewing some of the history associated with quaternions, and the development of octonions. The short section on composition algebras reveals that quaternions are rather special, and informs us why Hamilton could not identify an algebra based upon the hyper-complex number $z = s + ai + bj$.

We then examine various quaternion products to discover their rotational properties. This begins with two orthogonal quaternions, and moves towards the general case of using qpq^{-1} where q is a unit-norm quaternion, and p is a pure quaternion.

Two techniques are covered to express a quaternion product as a matrix, which in turn encode the eigenvector and eigenvalue. Finally, we examine how quaternions can be interpolated.

We continue to represent a quaternion as an ordered pair, with italic, lower-case letters to represent quaternions, and bold lower-case letters to represent vectors.

8.2 Some History

Hamilton invented quaternions in October 1843, and by December of the same year, his friend, Irish mathematician, John Thomas Graves (1806–1870), had invented *octaves*, which would eventually be called *octonions*. The British mathematician, Arthur Cayley (1821–1895), had also been intrigued by Hamilton's quaternions, and independently invented octonions in 1845. Octonions eventually became known as *Cayley numbers* rather than *octaves*, simply because Graves did not publish his results until 1848—three years after Cayley!

Just as quaternions can be defined in terms of ordered pairs of complex numbers, the octaves, or octonions, can be defined as ordered pairs of quaternions.

J. Vince, *Quaternions for Computer Graphics*,
https://doi.org/10.1007/978-1-4471-7509-4_8

8.2.1 Composition Algebras

When a specific law forms the basis of an algebra, it is called a *composition algebra*. For example, we know that in ordinary arithmetic

$$a^2b^2 = (ab)^2, \quad a, b \in \mathbb{R}.$$

For example:

$$3^2 4^2 = (3 \times 4)^2$$

where a square law is the composition law.

We discovered in Chap. 3 that for two complex numbers:

$$|z_1||z_2| = |z_1 z_2|, \quad z_1, z_2 \in \mathbb{C}$$
$$|z_1|^2|z_2|^2 = |z_1 z_2|^2.$$

For example, given

$$z_1 = a_1 + b_1 i$$
$$z_2 = a_2 + b_2 i$$

then

$$\left(a_1^2 + b_1^2\right)\left(a_2^2 + b_2^2\right) = (a_1a_2 - b_1b_2)^2 + (a_1b_2 + a_2b_1)^2$$

because:

$$(a_1^2 + b_1^2)(a_2^2 + b_2^2) = a_1^2a_2^2 + a_1^2b_2^2 + a_2^2b_1^2 + b_1^2b_2^2$$
$$(a_1a_2 - b_1b_2)^2 = a_1^2a_2^2 - 2a_1a_2b_1b_2 + b_1^2b_2^2$$
$$(a_1b_2 + a_2b_1)^2 = a_1^2b_2^2 + 2a_1a_2b_1b_2 + a_2^2b_1^2$$
$$(a_1a_2 - b_1b_2)^2 + (a_1b_2 + a_2b_1)^2 = a_1^2a_2^2 + b_1^2b_2^2 + a_1^2b_2^2 + a_2^2b_1^2$$

which is a two-square law.

In Chap. 6 we noted that for two quaternions:

$$|q_a|^2|q_b|^2 = |q_a q_b|^2, \quad q_a, q_b \in \mathbb{H}.$$

For example, given

$$q_a = [s_a, \; x_a\mathbf{i} + y_a\mathbf{j} + z_a\mathbf{k}]$$
$$q_b = [s_b, \; x_b\mathbf{i} + y_b\mathbf{j} + z_b\mathbf{k}]$$

then

$$\begin{aligned}\left(s_a^2 + x_a^2 + y_a^2 + z_a^2\right)\left(s_b^2 + x_b^2 + y_b^2 + z_b^2\right) &= (s_a s_b - x_a x_b - y_a y_b - z_a z_b)^2 \\ &+ (s_a x_b + s_b x_a + y_a z_b - y_b z_a)^2 \\ &+ (s_a y_b + s_b y_a + z_a x_b - z_b x_a)^2 \\ &+ (s_a z_b + s_b z_a + x_a y_b - x_b y_a)^2\end{aligned}$$

which is a four-square law.

In addition to complex numbers, quaternions occupy a central place in mathematical systems, and today there are four such composition algebras: real $\mathbb{R}$, complex $\mathbb{C}$, quaternion $\mathbb{H}$, and octonion $\mathbb{O}$ that obey an n-square identity used to compute their norms. In 1898 the German mathematician Adolf Hurwitz (1859–1919), proved that the product of the sum of n squares by the sum of n squares is the sum of n squares only when n is equal to 1, 2, 4 and 8, which are represented by the reals, complex numbers, quaternions and octonions. This is known as *Hurwitz's Theorem* or the *1, 2, 4, 8 Theorem*. No other system is possible, which shows how important quaternions are within the realm of mathematics. Consequently, Hamilton's search for a system of triples was futile, because there is no three-square identity.

Now let's investigate how quaternions are used to rotate vectors about an arbitrary axis.

8.3 Quaternion Products

A quaternion q is the union of a scalar s and a vector $\mathbf{v}$:

$$q = [s,\ \mathbf{v}], \qquad s \in \mathbb{R}, \quad \mathbf{v} \in \mathbb{R}^3.$$

If we express $\mathbf{v}$ in terms of its components, we have

$$q = [s,\ x\mathbf{i} + y\mathbf{j} + z\mathbf{k}], \quad s, x, y, z \in \mathbb{R}.$$

When two such quaternions are multiplied together, we obtain a third quaternion:

$$\begin{aligned} q_a &= [s_a,\ \mathbf{v}_a] \\ q_b &= [s_b,\ \mathbf{v}_b] \\ q_a q_b &= [s_a,\ \mathbf{v}_a][s_b,\ \mathbf{v}_b] \\ &= [s_a s_b - \mathbf{v}_a \cdot \mathbf{v}_b,\ s_a \mathbf{v}_b + s_b \mathbf{v}_a + \mathbf{v}_a \times \mathbf{v}_b]. \end{aligned}$$

Naturally, if s_a or s_b are zero, as in the case of a pure quaternion, the product is simplified. Therefore, in future I will omit any zero terms, to simplify the algebra.

Hamilton had hoped that a quaternion could be used like a complex rotor, where we saw in Chap. 4 that

$$\mathbf{R}_\theta = \cos\theta + i\sin\theta$$

rotates a complex number by θ. Could a unit-norm quaternion q be used to rotate a vector stored as a pure quaternion p? Well yes, but only as a special case. To understand this, let's construct the product of a unit-norm quaternion q and a pure quaternion p. The unit-norm quaternion q is defined as

$$\begin{aligned} q &= [s,\ \lambda\hat{\mathbf{v}}], \quad s, \lambda \in \mathbb{R}, \quad \hat{\mathbf{v}} \in \mathbb{R}^3, \\ \|\hat{\mathbf{v}}\| &= 1 \\ s^2 + \lambda^2 &= 1 \end{aligned} \tag{8.1}$$

and the pure quaternion p stores the vector $\mathbf{p}$ to be rotated:

$$p = [0,\ \mathbf{p}], \quad \mathbf{p} \in \mathbb{R}^3.$$

Let's compute the product $p' = qp$ and examine the vector part of p' to see if $\mathbf{p}$ is rotated:

$$\begin{aligned} q &= [s,\ \lambda\hat{\mathbf{v}}] \\ p &= [0,\ \mathbf{p}] \\ p' &= qp \\ &= [s,\ \lambda\hat{\mathbf{v}}][0,\ \mathbf{p}] \\ &= [-\lambda\hat{\mathbf{v}} \cdot \mathbf{p},\ s\mathbf{p} + \lambda\hat{\mathbf{v}} \times \mathbf{p}]. \end{aligned} \tag{8.2}$$

We can see from (8.2) that the result is a general quaternion with a scalar and a vector component.

8.3.1 Special Case

The 'special case' referred to above is that $\hat{\mathbf{v}}$ must be perpendicular to $\mathbf{p}$, which makes the dot product term $-\lambda\hat{\mathbf{v}} \cdot \mathbf{p}$ in (8.2) vanish, and we are left with the pure quaternion:

$$p' = [0,\ s\mathbf{p} + \lambda\hat{\mathbf{v}} \times \mathbf{p}]. \tag{8.3}$$

Figure 8.1 illustrates this scenario, where $\mathbf{p}$ is perpendicular to $\hat{\mathbf{v}}$, and $\hat{\mathbf{v}} \times \mathbf{p}$ is perpendicular to the plane containing $\mathbf{p}$ and $\hat{\mathbf{v}}$.

Now as $\hat{\mathbf{v}}$ is a unit vector, $\|\mathbf{p}\| = \|\hat{\mathbf{v}} \times \mathbf{p}\|$, which means that we have two orthogonal vectors, i.e. $\mathbf{p}$ and $\hat{\mathbf{v}} \times \mathbf{p}$, with the same length. Therefore, to rotate $\mathbf{p}$ about $\hat{\mathbf{v}}$, all that we have to do is make $s = \cos\theta$ and $\lambda = \sin\theta$ in (8.3):

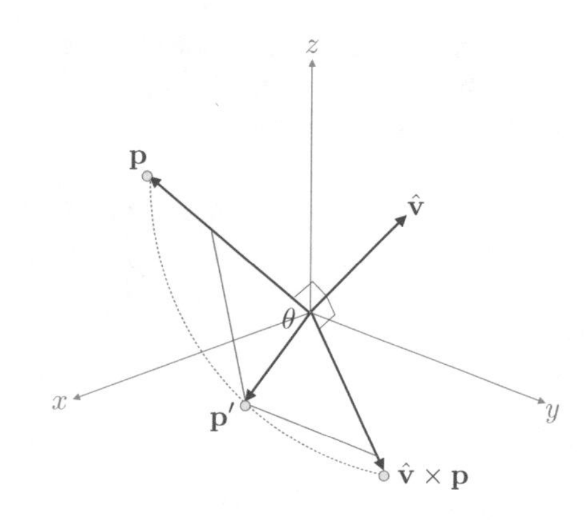

Fig. 8.1 Three orthogonal vectors $\mathbf{p}$, $\hat{\mathbf{v}}$ and $\hat{\mathbf{v}} \times \mathbf{p}$

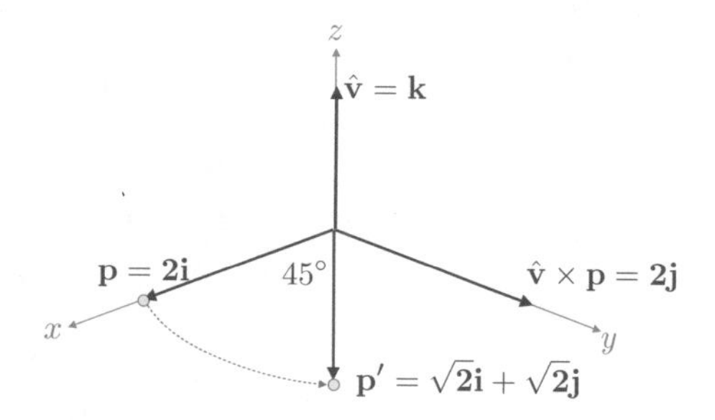

Fig. 8.2 The vector $2\mathbf{i}$ is rotated 45° by the quaternion $q = \left[\frac{\sqrt{2}}{2}, \frac{\sqrt{2}}{2}\mathbf{k}\right]$

$$\begin{aligned} p' &= [0, \ \mathbf{p}'] \\ &= [0, \ \cos\theta\mathbf{p} + \sin\theta\hat{\mathbf{v}} \times \mathbf{p}]. \end{aligned}$$

For example, to rotate a vector about the z-axis, q's vector $\hat{\mathbf{v}}$ must be aligned with the z-axis as shown in Fig. 8.2. If we make the angle of rotation $\theta = 45°$ then

$$\begin{aligned} q &= [s, \ \lambda\hat{\mathbf{v}}] \\ &= [\cos\theta, \ \sin\theta\mathbf{k}] \\ &= \left[\frac{\sqrt{2}}{2}, \ \frac{\sqrt{2}}{2}\mathbf{k}\right] \\ &= \frac{\sqrt{2}}{2}[1, \ \mathbf{k}] \end{aligned}$$

and if the vector to be rotated is $\mathbf{p} = 2\mathbf{i}$, then

$$\begin{aligned} p &= [0, \ \mathbf{p}] \\ &= [0, \ 2\mathbf{i}] \\ &= 2[0, \ \mathbf{i}]. \end{aligned}$$

There are now four product combinations worth exploring: $qp, pq, q^{-1}p$ and pq^{-1}. It's not worth considering qp^{-1} and $p^{-1}q$ as p^{-1} simply reverses the direction of $\mathbf{p}$. Let's start with qp:

$$\begin{aligned}
q &= \tfrac{\sqrt{2}}{2}[1,\ \mathbf{k}] \\
p &= 2[0,\ \mathbf{i}] \\
p' &= qp \\
&= \sqrt{2}[1,\ \mathbf{k}][0,\ \mathbf{i}] \\
&= \sqrt{2}[0,\ \mathbf{i}+\mathbf{j}]
\end{aligned}$$

and $\mathbf{p}$ has been rotated 45° to $\mathbf{p}' = \sqrt{2}\mathbf{i} + \sqrt{2}\mathbf{j}$.

Next, pq:

$$\begin{aligned}
p &= 2[0,\ \mathbf{i}] \\
q &= \tfrac{\sqrt{2}}{2}[1,\ \mathbf{k}] \\
p' &= pq \\
&= \sqrt{2}[0,\ \mathbf{i}][1,\ \mathbf{k}] \\
&= \sqrt{2}[0,\ \mathbf{i}+\mathbf{i}\times\mathbf{k}] \\
&= \sqrt{2}[0,\ \mathbf{i}-\mathbf{j}]
\end{aligned}$$

and $\mathbf{p}$ has been rotated $-45°$ to $\mathbf{p}' = \sqrt{2}\mathbf{i} - \sqrt{2}\mathbf{j}$.

Next, $q^{-1}p$, and as q is a unit-norm quaternion, $q^{-1} = q^*$:

$$\begin{aligned}
q &= \tfrac{\sqrt{2}}{2}[1,\ \mathbf{k}] \\
q^{-1} &= \tfrac{\sqrt{2}}{2}[1,\ -\mathbf{k}] \\
p &= 2[0,\ \mathbf{i}] \\
p' &= q^{-1}p \\
&= \sqrt{2}[1,\ -\mathbf{k}][0,\ \mathbf{i}] \\
&= \sqrt{2}[0,\ \mathbf{i}-\mathbf{k}\times\mathbf{i}] \\
&= \sqrt{2}[0,\ \mathbf{i}-\mathbf{j}]
\end{aligned}$$

and $\mathbf{p}$ has been rotated $-45°$ to $\mathbf{p}' = \sqrt{2}\mathbf{i} - \sqrt{2}\mathbf{j}$.

Finally, pq^{-1}:

$$
\begin{aligned}
p &= 2[0,\ \mathbf{i}] \\
q &= \tfrac{\sqrt{2}}{2}[1,\ \mathbf{k}] \\
q^{-1} &= \tfrac{\sqrt{2}}{2}[1,\ -\mathbf{k}] \\
p' &= pq^{-1} \\
&= \sqrt{2}[0,\ \mathbf{i}][1,\ -\mathbf{k}] \\
&= \sqrt{2}[0,\ \mathbf{i} - \mathbf{i} \times \mathbf{k}] \\
&= \sqrt{2}[0,\ \mathbf{i} + \mathbf{j}]
\end{aligned}
$$

and $\mathbf{p}$ has been rotated 45° to $\mathbf{p}' = \sqrt{2}\mathbf{i} + \sqrt{2}\mathbf{j}$. Thus, for orthogonal quaternions, θ is the angle of rotation, then

$$
\begin{aligned}
qp &= pq^{-1} \\
pq &= q^{-1}p.
\end{aligned}
$$

Before moving on, let's see what happens to the product qp when $\theta = 180°$:

$$
\begin{aligned}
q &= [\cos\theta,\ \sin\theta\mathbf{k}] \\
&= [-1,\ \mathbf{0}] \\
p &= 2[0,\ \mathbf{i}] \\
p' &= qp \\
&= 2[-1,\ \mathbf{0}][0,\ \mathbf{i}] \\
&= 2[0,\ -\mathbf{i} + \mathbf{0} \times \mathbf{i}] \\
&= [0,\ -2\mathbf{i}]
\end{aligned}
$$

and $\mathbf{p}$ has been rotated 180° to $\mathbf{p}' = -2\mathbf{i}$.

Note that in all the above products, the vector has not been scaled during the rotation. This is because q is a unit-norm quaternion.

Now let's see what happens if we change the angle between $\hat{\mathbf{v}}$ and $\mathbf{p}$. Let's reduce the angle to 45° and retain q's unit vector, as shown in Fig. 8.3, such that $\hat{\mathbf{v}}$ is directed along the z-axis, and $\mathbf{p} = \mathbf{i} + \mathbf{k}$. Therefore,

$$
\begin{aligned}
\hat{\mathbf{v}} &= \mathbf{k} \\
q &= \left[\cos\theta,\ \sin\theta\hat{\mathbf{v}}\right] \\
p &= [0,\ \mathbf{p}].
\end{aligned}
$$

This time we must include the dot product term $-\sin\theta\hat{\mathbf{v}} \cdot \mathbf{p}$, as it is no longer zero:

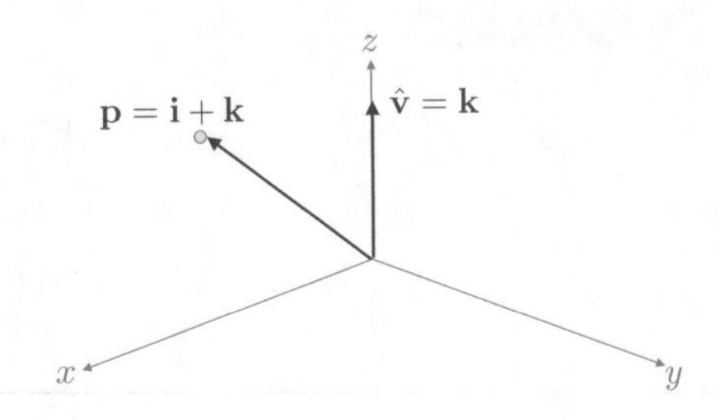

Fig. 8.3 Rotating the vector $\mathbf{p} = \mathbf{i} + \mathbf{k}$ by the quaternion $q = \big[\cos\theta,\ \sin\theta\hat{\mathbf{v}}\big]$

$$\begin{aligned} q &= [\cos\theta,\ \sin\theta\hat{\mathbf{v}}] \\ p &= [0,\ \mathbf{p}] \\ p' &= qp \\ &= [\cos\theta,\ \sin\theta\hat{\mathbf{v}}][0,\ \mathbf{p}] \\ &= [-\sin\theta\hat{\mathbf{v}}\cdot\mathbf{p},\ \cos\theta\mathbf{p} + \sin\theta\hat{\mathbf{v}}\times\mathbf{p}]. \end{aligned} \tag{8.4}$$

Substituting $\hat{\mathbf{v}}$, $\mathbf{p}$ and $\theta = 45°$ in (8.4), we have

$$\begin{aligned} \hat{\mathbf{v}} &= \mathbf{k} \\ \mathbf{p} &= \mathbf{i} + \mathbf{k} \\ p' &= \Big[-\tfrac{\sqrt{2}}{2}\mathbf{k}\cdot(\mathbf{i}+\mathbf{k}),\ \tfrac{\sqrt{2}}{2}(\mathbf{i}+\mathbf{k}) + \tfrac{\sqrt{2}}{2}\mathbf{k}\times(\mathbf{i}+\mathbf{k})\Big] \\ &= \Big[-\tfrac{\sqrt{2}}{2},\ \tfrac{\sqrt{2}}{2}\mathbf{i} + \tfrac{\sqrt{2}}{2}\mathbf{k} + \tfrac{\sqrt{2}}{2}\mathbf{j}\Big] \\ &= \tfrac{\sqrt{2}}{2}[-1,\ \mathbf{i}+\mathbf{j}+\mathbf{k}] \end{aligned} \tag{8.5}$$

which, unfortunately, is no longer a pure quaternion. Multiplying the vector by a non-orthogonal quaternion has converted some of the vector information into the quaternion's scalar component.

8.3.2 General Case

Not to worry. Could it be that an inverse quaternion reverses the operation? Let's see what happens if we post-multiply qp by q^{-1}.

Given

$$q = [\cos\theta,\ \sin\theta\mathbf{k}]$$

then

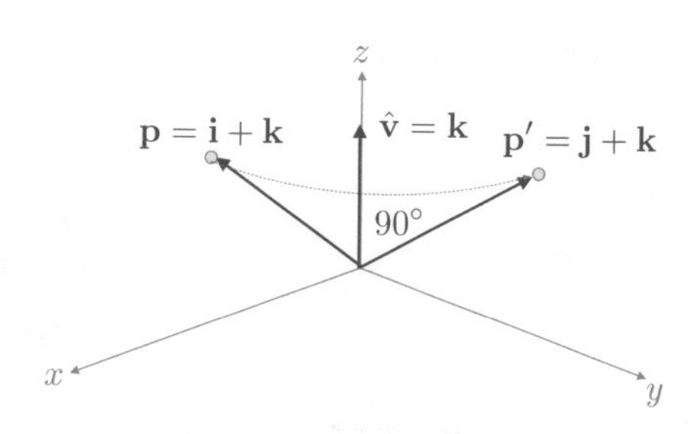

Fig. 8.4 The vector $\mathbf{i}+\mathbf{k}$ is rotated 90° to $\mathbf{j}+\mathbf{k}$

$$\begin{aligned}
q^{-1} &= [\cos\theta,\ -\sin\theta\mathbf{k}] \\
&= \left[\tfrac{\sqrt{2}}{2},\ -\tfrac{\sqrt{2}}{2}\mathbf{k}\right] \\
&= \tfrac{\sqrt{2}}{2}[1,\ -\mathbf{k}].
\end{aligned}$$

Therefore, post-multiplying (8.5) by q^{-1} we have:

$$\begin{aligned}
qp &= \tfrac{\sqrt{2}}{2}[-1,\ \mathbf{i}+\mathbf{j}+\mathbf{k}] \\
q^{-1} &= \tfrac{\sqrt{2}}{2}[1,\ -\mathbf{k}] \\
qpq^{-1} &= \tfrac{\sqrt{2}}{2}[-1,\ \mathbf{i}+\mathbf{j}+\mathbf{k}]\tfrac{\sqrt{2}}{2}[1,\ -\mathbf{k}] \\
&= \tfrac{1}{2}[-1,\ \mathbf{i}+\mathbf{j}+\mathbf{k}][1,\ -\mathbf{k}] \\
&= \tfrac{1}{2}[-1+1,\ \mathbf{k}+\mathbf{i}+\mathbf{j}+\mathbf{k}+(\mathbf{i}+\mathbf{j}+\mathbf{k})\times-\mathbf{k})] \\
&= \tfrac{1}{2}[0,\ \mathbf{i}+\mathbf{j}+2\mathbf{k}-\mathbf{i}+\mathbf{j}] \\
&= [0,\ \mathbf{j}+\mathbf{k}].
\end{aligned} \tag{8.6}$$

Equation (8.6) is a pure quaternion, with a norm of $\sqrt{2}$, which is the same as $\mathbf{p}$. However, the vector has been rotated 90° rather than 45°, twice the desired angle, as shown in Fig. 8.4.

If this 'sandwiching' of the vector in the form of a pure quaternion by q and q^{-1} is correct, it suggests that increasing θ to 90° should rotate $\mathbf{p}=\mathbf{i}+\mathbf{k}$ by 180° to $-\mathbf{i}+\mathbf{k}$. Let's try this.

Let $\theta = 90°$, therefore,

$$\begin{aligned}
q &= [\cos 90°,\ \sin 90°\mathbf{k}] \\
&= [0,\ \mathbf{k}] \\
p &= [0,\ \mathbf{i}+\mathbf{k}] \\
qp &= [0,\ \mathbf{k}][0,\ \mathbf{i}+\mathbf{k}] \\
&= [-1,\ \mathbf{k}\times(\mathbf{i}+\mathbf{k})] \\
&= [-1,\ \mathbf{j}].
\end{aligned}$$

Next, we post-multiply qp by q^{-1}:

$$
\begin{aligned}
q^{-1} &= [0,\ -\mathbf{k}] \\
qpq^{-1} &= [-1,\ \mathbf{j}][0,\ -\mathbf{k}] \\
&= [0,\ \mathbf{k} + (\mathbf{j} \times -\mathbf{k})] \\
&= [0,\ -\mathbf{i} + \mathbf{k}]
\end{aligned}
$$

which confirms our prediction and suggests that qpq^{-1} works.

8.3.3 Double Angle

Now let's show how this double angle arises. We begin by defining a unit-norm quaternion q:

$$q = [s,\ \lambda\hat{\mathbf{v}}]$$

where $s^2 + \lambda^2 = 1$. The vector $\mathbf{p}$ to be rotated is encoded as a pure quaternion:

$$p = [0,\ \mathbf{p}]$$

and the inverse quaternion q^{-1} is

$$q^{-1} = [s,\ -\lambda\hat{\mathbf{v}}].$$

Therefore, the product qpq^{-1} is

$$
\begin{aligned}
qpq^{-1} &= \left[s,\ \lambda\hat{\mathbf{v}}\right][0,\ \mathbf{p}][s,\ -\lambda\hat{\mathbf{v}}] \\
&= \left[-\lambda\hat{\mathbf{v}} \cdot \mathbf{p},\ s\mathbf{p} + \lambda\hat{\mathbf{v}} \times \mathbf{p}\right][s,\ -\lambda\hat{\mathbf{v}}] \\
&= \big[-\lambda s\hat{\mathbf{v}} \cdot \mathbf{p} + \lambda s\mathbf{p} \cdot \hat{\mathbf{v}} + \lambda^2(\hat{\mathbf{v}} \times \mathbf{p}) \cdot \hat{\mathbf{v}}, \\
&\quad + \lambda^2(\hat{\mathbf{v}} \cdot \mathbf{p})\hat{\mathbf{v}} + s^2\mathbf{p} + \lambda s\hat{\mathbf{v}} \times \mathbf{p} \\
&\quad - \lambda s\mathbf{p} \times \hat{\mathbf{v}} - \lambda^2(\hat{\mathbf{v}} \times \mathbf{p}) \times \hat{\mathbf{v}}\big] \\
&= \left[\lambda^2(\hat{\mathbf{v}} \times \mathbf{p}) \cdot \hat{\mathbf{v}},\ \lambda^2(\hat{\mathbf{v}} \cdot \mathbf{p})\hat{\mathbf{v}} + s^2\mathbf{p} + 2\lambda s\hat{\mathbf{v}} \times \mathbf{p} - \lambda^2(\hat{\mathbf{v}} \times \mathbf{p}) \times \hat{\mathbf{v}}\right].
\end{aligned}
$$

Note that

$$(\hat{\mathbf{v}} \times \mathbf{p}) \cdot \hat{\mathbf{v}} = 0$$

and

$$(\hat{\mathbf{v}} \times \mathbf{p}) \times \hat{\mathbf{v}} = (\hat{\mathbf{v}} \cdot \hat{\mathbf{v}})\mathbf{p} - (\mathbf{p} \cdot \hat{\mathbf{v}})\hat{\mathbf{v}} = \mathbf{p} - (\mathbf{p} \cdot \hat{\mathbf{v}})\hat{\mathbf{v}}.$$

Therefore,

$$\begin{aligned}
qpq^{-1} &= \left[0,\ \lambda^2\left(\hat{\mathbf{v}}\cdot\mathbf{p}\right)\hat{\mathbf{v}} + s^2\mathbf{p} + 2\lambda s\hat{\mathbf{v}}\times\mathbf{p} - \lambda^2\mathbf{p} + \lambda^2\left(\mathbf{p}\cdot\hat{\mathbf{v}}\right)\hat{\mathbf{v}}\right] \\
&= \left[0,\ 2\lambda^2\left(\hat{\mathbf{v}}\cdot\mathbf{p}\right)\hat{\mathbf{v}} + \left(s^2-\lambda^2\right)\mathbf{p} + 2\lambda s\hat{\mathbf{v}}\times\mathbf{p}\right].
\end{aligned} \tag{8.7}$$

Clearly, (8.7) is a pure quaternion as the scalar component is zero. However, it is not obvious where the angle doubling comes from. But look what happens when we make $s = \cos\theta$ and $\lambda = \sin\theta$:

$$\begin{aligned}
qpq^{-1} &= \left[0,\ 2\sin^2\theta\left(\hat{\mathbf{v}}\cdot\mathbf{p}\right)\hat{\mathbf{v}} + \left(\cos^2\theta - \sin^2\theta\right)\mathbf{p} + 2\sin\theta\cos\theta\hat{\mathbf{v}}\times\mathbf{p}\right] \\
&= \left[0,\ (1-\cos(2\theta))\left(\hat{\mathbf{v}}\cdot\mathbf{p}\right)\hat{\mathbf{v}} + \cos(2\theta)\mathbf{p} + \sin(2\theta)\hat{\mathbf{v}}\times\mathbf{p}\right].
\end{aligned}$$

The double-angle trigonometric terms emerge! Now, if we want this product to actually rotate the vector by θ, then we must build this in from the outset by halving θ in q:

$$q = \left[\cos\left(\tfrac{\theta}{2}\right),\ \sin\left(\tfrac{\theta}{2}\right)\hat{\mathbf{v}}\right] \tag{8.8}$$

which makes

$$qpq^{-1} = \left[0,\ (1-\cos\theta)\left(\hat{\mathbf{v}}\cdot\mathbf{p}\right)\hat{\mathbf{v}} + \cos\theta\mathbf{p} + \sin\theta\hat{\mathbf{v}}\times\mathbf{p}\right]. \tag{8.9}$$

The product qpq^{-1} was discovered by Hamilton who failed to publish the result. Cayley, also discovered the product and published the result in 1845 [1]. However, Altmann notes that "in Cayley's collected papers he concedes priority to Hamilton." [2], which was a nice gesture. However, the person who had recognised the importance of the half-angle parameters in (8.8) before Hamilton and Cayley was Rodrigues—who published a solution that was not seen by Hamilton, but apparently, was seen by Cayley.

Let's test (8.9) using the previous example where we rotated a vector $\mathbf{p} = \mathbf{i} + \mathbf{k}$, $\theta = 90°$ about the quaternion's vector $\hat{\mathbf{v}} = \mathbf{k}$.

$$\begin{aligned}
qpq^{-1} &= \left[0,\ (1-\cos\theta)(\hat{\mathbf{v}}\cdot\mathbf{p})\hat{\mathbf{v}} + \cos\theta\mathbf{p} + \sin\theta\hat{\mathbf{v}}\times\mathbf{p}\right] \\
&= \left[0,\ (\hat{\mathbf{v}}\cdot\mathbf{p})\hat{\mathbf{v}} + \hat{\mathbf{v}}\times\mathbf{p}\right] \\
&= [0,\ (\mathbf{k}\cdot(\mathbf{i}+\mathbf{k}))\mathbf{k} + \mathbf{j}] \\
&= [0,\ \mathbf{j}+\mathbf{k}]
\end{aligned}$$

which agrees with (8.6). Thus, when a unit-norm quaternion takes the form

$$q = \left[\cos\left(\tfrac{\theta}{2}\right),\ \sin\left(\tfrac{\theta}{2}\right)\hat{\mathbf{v}}\right]$$

and a pure quaternion storing a vector to be rotated takes the form

$$p = [0,\ \mathbf{p}]$$

the pure quaternion

$$p' = qpq^{-1}$$

stores the rotated vector $\mathbf{p}'$. Let's show why this product preserves the magnitude of the rotated vector.

$$\begin{aligned}|p'| &= |qp||q^{-1}| \\ &= |q||p||q^{-1}| \\ &= |q|^2|p|\end{aligned}$$

and if q is a unit-norm quaternion, $|q| = 1$, then $|p'| = |p|$.

You may be wondering what happens if the product is reversed to $q^{-1}pq$? A guess would suggest that the rotation sequence is reversed, but let's see what an algebraic analysis confirms.

$$\begin{aligned}q^{-1}pq &= [s,\ -\lambda\hat{\mathbf{v}}][0,\ \mathbf{p}][s,\ \lambda\hat{\mathbf{v}}] \\ &= [\lambda\hat{\mathbf{v}}\cdot\mathbf{p},\ s\mathbf{p} - \lambda\hat{\mathbf{v}}\times\mathbf{p}][s,\ \lambda\hat{\mathbf{v}}] \\ &= \big[\lambda s\hat{\mathbf{v}}\cdot\mathbf{p} - \lambda s\mathbf{p}\cdot\hat{\mathbf{v}}, \\ &\quad \lambda^2\hat{\mathbf{v}}\times\mathbf{p}\cdot\hat{\mathbf{v}} + \lambda^2\hat{\mathbf{v}}\cdot\mathbf{p}\hat{\mathbf{v}} + s^2\mathbf{p} - \lambda s\hat{\mathbf{v}}\times\mathbf{p} + \lambda s\mathbf{p}\times\hat{\mathbf{v}} - \lambda^2\hat{\mathbf{v}}\times\mathbf{p}\times\hat{\mathbf{v}}\big] \\ &= \big[\lambda^2(\hat{\mathbf{v}}\times\mathbf{p})\cdot\hat{\mathbf{v}},\ \lambda^2(\hat{\mathbf{v}}\cdot\mathbf{p})\hat{\mathbf{v}} + s^2\mathbf{p} - 2\lambda s\hat{\mathbf{v}}\times\mathbf{p} - \lambda^2(\hat{\mathbf{v}}\times\mathbf{p})\times\hat{\mathbf{v}}\big].\end{aligned}$$

Once again

$$(\hat{\mathbf{v}}\times\mathbf{p})\cdot\hat{\mathbf{v}} = 0$$

and

$$(\hat{\mathbf{v}}\times\mathbf{p})\times\hat{\mathbf{v}} = \mathbf{p} - (\mathbf{p}\cdot\hat{\mathbf{v}})\hat{\mathbf{v}}.$$

Therefore,

$$\begin{aligned}q^{-1}pq &= \big[0,\ \lambda^2(\hat{\mathbf{v}}\cdot\mathbf{p})\hat{\mathbf{v}} + s^2\mathbf{p} - 2\lambda s\hat{\mathbf{v}}\times\mathbf{p} - \lambda^2\mathbf{p} + \lambda^2(\mathbf{p}\cdot\hat{\mathbf{v}})\hat{\mathbf{v}}\big] \\ &= \big[0,\ 2\lambda^2(\hat{\mathbf{v}}\cdot\mathbf{p})\hat{\mathbf{v}} + (s^2 - \lambda^2)\mathbf{p} - 2\lambda s\hat{\mathbf{v}}\times\mathbf{p}\big].\end{aligned}$$

Again, let's make $s = \cos\theta$ and $\lambda = \sin\theta$:

$$q^{-1}pq = \big[0,\ (1-\cos(2\theta))(\hat{\mathbf{v}}\cdot\mathbf{p})\hat{\mathbf{v}} + \cos(2\theta)\mathbf{p} - \sin(2\theta)\hat{\mathbf{v}}\times\mathbf{p}\big]$$

and the only thing that has changed from qpq^{-1} is the sign of the cross-product term, which reverses the direction of its vector. However, we must remember to compensate for the angle-doubling by halving θ:

$$q^{-1}pq = \big[0,\ (1-\cos\theta)(\hat{\mathbf{v}}\cdot\mathbf{p})\hat{\mathbf{v}} + \cos\theta\mathbf{p} - \sin\theta\hat{\mathbf{v}}\times\mathbf{p}\big]. \tag{8.10}$$

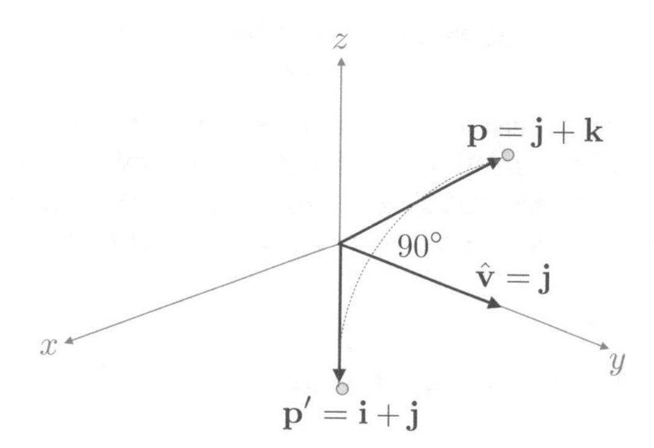

Fig. 8.5 The point $P(0, 1, 1)$ is rotated 90° to $P'(1, 1, 0)$ about the y-axis

Let's see what happens when we employ (8.10) to rotate $\mathbf{p} = \mathbf{i} + \mathbf{k}$, 90° about the quaternion's vector $\hat{\mathbf{v}} = \mathbf{k}$:

$$\begin{aligned} q^{-1}pq &= [0,\ (\mathbf{k} \cdot (\mathbf{i} + \mathbf{k})\mathbf{k}) - \mathbf{j}] \\ &= [0,\ -\mathbf{j} + \mathbf{k}] \end{aligned}$$

which has rotated $\mathbf{p}$ clockwise 90° about the quaternion's vector. Therefore, the rotor qpq^{-1} rotates a vector counter-clockwise, and $q^{-1}pq$ rotates a vector clockwise:

$$\begin{aligned} qpq^{-1} &= \big[0,\ (1 - \cos\theta)(\hat{\mathbf{v}} \cdot \mathbf{p})\hat{\mathbf{v}} + \cos\theta\mathbf{p} + \sin\theta\hat{\mathbf{v}} \times \mathbf{p}\big] \\ q^{-1}pq &= \big[0,\ (1 - \cos\theta)(\hat{\mathbf{v}} \cdot \mathbf{p})\hat{\mathbf{v}} + \cos\theta\mathbf{p} - \sin\theta\hat{\mathbf{v}} \times \mathbf{p}\big]. \end{aligned}$$

Let's compute another example. Consider the point $P(0, 1, 1)$ in Fig. 8.5 which is to be rotated 90° about the y-axis. We can see that the rotated point P' has the coordinates $(1, 1, 0)$ which we will confirm algebraically. The point P is represented by its position vector $\mathbf{p}$ in the pure quaternion

$$p = [0,\ \mathbf{p}].$$

The axis of rotation is $\hat{\mathbf{v}} = \mathbf{j}$, and the vector to be rotated is $\mathbf{p} = \mathbf{j} + \mathbf{k}$. Therefore,

$$\begin{aligned} qpq^{-1} &= \big[0,\ (1 - \cos\theta)(\hat{\mathbf{v}} \cdot \mathbf{p})\hat{\mathbf{v}} + \cos\theta\mathbf{p} + \sin\theta\hat{\mathbf{v}} \times \mathbf{p}\big] \\ &= [0,\ \mathbf{j} \cdot (\mathbf{j} + \mathbf{k})\,\mathbf{j} + \mathbf{j} \times (\mathbf{j} + \mathbf{k})] \\ &= [0,\ \mathbf{i} + \mathbf{j}] \end{aligned}$$

and confirms that P is indeed rotated to $(1, 1, 0)$.

Now let's explore how this product is represented in matrix form.

8.4 Quaternions in Matrix Form

Having discovered a vector equation to represent qpq^{-1}, let's continue and convert it into a matrix. We will explore two methods: the first is a simple vectorial method which translates the vector equation representing qpq^{-1} directly into matrix form. The second method uses matrix algebra to develop a rather cunning solution.

8.4.1 Vector Method

For the vector method it is convenient to describe the unit-norm quaternion as

$$\begin{aligned} q &= [s,\ \mathbf{v}] \\ &= [s,\ x\mathbf{i} + y\mathbf{j} + z\mathbf{k}] \end{aligned}$$

where

$$s^2 + \|\mathbf{v}\|^2 = 1$$

and the pure quaternion as

$$\begin{aligned} p &= [0,\ \mathbf{p}] \\ &= [0,\ x_p\mathbf{i} + y_p\mathbf{j} + z_p\mathbf{k}]. \end{aligned}$$

A simple way to compute qpq^{-1} is to use (8.9) and substitute $\|\mathbf{v}\|$ for λ:

$$\begin{aligned} qpq^{-1} &= \left[0,\ 2\lambda^2\left(\hat{\mathbf{v}} \cdot \mathbf{p}\right)\hat{\mathbf{v}} + (s^2 - \lambda^2)\mathbf{p} + 2\lambda s\hat{\mathbf{v}} \times \mathbf{p}\right] \\ &= \left[0,\ 2\|\mathbf{v}\|^2\left(\hat{\mathbf{v}} \cdot \mathbf{p}\right)\hat{\mathbf{v}} + (s^2 - \|\mathbf{v}\|^2)\mathbf{p} + 2\|\mathbf{v}\|s\hat{\mathbf{v}} \times \mathbf{p}\right]. \end{aligned}$$

Next, we substitute $\mathbf{v}$ for $\|\mathbf{v}\|\hat{\mathbf{v}}$:

$$qpq^{-1} = \left[0,\ 2\left(\mathbf{v} \cdot \mathbf{p}\right)\mathbf{v} + \left(s^2 - \|\mathbf{v}\|^2\right)\mathbf{p} + 2s\mathbf{v} \times \mathbf{p}\right].$$

Finally, as we are working with unit-norm quaternions to prevent scaling

$$s^2 + \|\mathbf{v}\|^2 = 1$$

and

$$s^2 - \|\mathbf{v}\|^2 = 2s^2 - 1$$

therefore,

$$qpq^{-1} = \left[0,\ 2(\mathbf{v} \cdot \mathbf{p})\mathbf{v} + \left(2s^2 - 1\right)\mathbf{p} + 2s\mathbf{v} \times \mathbf{p}\right].$$

If we let $p' = qpq^{-1}$, which is a pure quaternion, we have

$$\begin{aligned}
p' &= qpq^{-1} \\
&= [0,\ \mathbf{p}'] \\
&= \left[0,\ 2(\mathbf{v}\cdot\mathbf{p})\mathbf{v} + \left(2s^2-1\right)\mathbf{p} + 2s\mathbf{v}\times\mathbf{p}\right] \\
\mathbf{p}' &= 2(\mathbf{v}\cdot\mathbf{p})\mathbf{v} + \left(2s^2-1\right)\mathbf{p} + 2s\mathbf{v}\times\mathbf{p}.
\end{aligned}$$

We are only interested in the rotated vector $\mathbf{p}'$ comprising the three terms $2(\mathbf{v}\cdot\mathbf{p})\mathbf{v}$, $\left(2s^2-1\right)\mathbf{p}$ and $2s\mathbf{v}\times\mathbf{p}$, which can be represented by three individual matrices and summed together.

$$\begin{aligned}
2(\mathbf{v}\cdot\mathbf{p})\mathbf{v} &= 2\left(xx_p + yy_p + zz_p\right)\left(x\mathbf{i} + y\mathbf{j} + z\mathbf{k}\right) \\
&= \begin{bmatrix} 2x^2 & 2xy & 2xz \\ 2xy & 2y^2 & 2yz \\ 2xz & 2yz & 2z^2 \end{bmatrix} \begin{bmatrix} x_p \\ y_p \\ z_p \end{bmatrix} \\
\left(2s^2-1\right)\mathbf{p} &= \left(2s^2-1\right)x_p\mathbf{i} + \left(2s^2-1\right)y_p\mathbf{j} + \left(2s^2-1\right)z_p\mathbf{k} \\
&= \begin{bmatrix} 2s^2-1 & 0 & 0 \\ 0 & 2s^2-1 & 0 \\ 0 & 0 & 2s^2-1 \end{bmatrix} \begin{bmatrix} x_p \\ y_p \\ z_p \end{bmatrix} \\
2s\mathbf{v}\times\mathbf{p} &= 2s\Big(\left(yz_p - zy_p\right)\mathbf{i} + \left(zx_p - xz_p\right)\mathbf{j} + \left(xy_p - yx_p\right)\mathbf{k}\Big) \\
&= \begin{bmatrix} 0 & -2sz & 2sy \\ 2sz & 0 & -2sx \\ -2sy & 2sx & 0 \end{bmatrix} \begin{bmatrix} x_p \\ y_p \\ z_p \end{bmatrix}.
\end{aligned}$$

Adding these matrices together:

$$\mathbf{p}' = \begin{bmatrix} 2\left(s^2+x^2\right)-1 & 2(xy-sz) & 2(xz+sy) \\ 2(xy+sz) & 2\left(s^2+y^2\right)-1 & 2(yz-sx) \\ 2(xz-sy) & 2(yz+sx) & 2\left(s^2+z^2\right)-1 \end{bmatrix} \begin{bmatrix} x_p \\ y_p \\ z_p \end{bmatrix} \tag{8.11}$$

or

$$\mathbf{p}' = \begin{bmatrix} 1-2\left(y^2+z^2\right) & 2(xy-sz) & 2(xz+sy) \\ 2(xy+sz) & 1-2\left(x^2+z^2\right) & 2(yz-sx) \\ 2(xz-sy) & 2(yz+sx) & 1-2\left(x^2+y^2\right) \end{bmatrix} \begin{bmatrix} x_p \\ y_p \\ z_p \end{bmatrix} \tag{8.12}$$

where

$$[0,\ \mathbf{p}'] = qpq^{-1}.$$

Now let's reverse the product. To compute the vector part of $q^{-1}pq$ all that we have to do is reverse the sign of $2s\mathbf{v}\times\mathbf{p}$:

$$\mathbf{p}' = \begin{bmatrix} 2\left(s^2+x^2\right)-1 & 2(xy+sz) & 2(xz-sy) \\ 2(xy-sz) & 2\left(s^2+y^2\right)-1 & 2(yz+sx) \\ 2(xz+sy) & 2(yz-sx) & 2\left(s^2+z^2\right)-1 \end{bmatrix} \begin{bmatrix} x_p \\ y_p \\ z_p \end{bmatrix} \quad (8.13)$$

or

$$\mathbf{p}' = \begin{bmatrix} 1-2\left(y^2+z^2\right) & 2(xy+sz) & 2(xz-sy) \\ 2(xy-sz) & 1-2\left(x^2+z^2\right) & 2(yz+sx) \\ 2(xz+sy) & 2(yz-sx) & 1-2\left(x^2+y^2\right) \end{bmatrix} \begin{bmatrix} x_p \\ y_p \\ z_p \end{bmatrix} \quad (8.14)$$

where

$$[0,\ \mathbf{p}'] = q^{-1}pq.$$

Observe that (8.13) is the transpose of (8.11), and (8.14) is the transpose of (8.12).

8.4.2 Matrix Method

The second method to derive (8.9) employs the matrix representing a quaternion product (6.14):

$$\begin{aligned} q_a &= [s_a,\ x_a\mathbf{i}+y_a\mathbf{j}+z_a\mathbf{k}] \\ q_b &= [s_b,\ x_b\mathbf{i}+y_b\mathbf{j}+z_b\mathbf{k}] \end{aligned}$$

and their product is

$$\begin{aligned} q_aq_b &= \big[s_a,\ x_a\mathbf{i}+y_a\mathbf{j}+z_a\mathbf{k}\big]\big[s_b,\ x_b\mathbf{i}+y_b\mathbf{j}+z_b\mathbf{k}\big] \\ &= \big[s_as_b-x_ax_b-y_ay_b-z_az_b, \\ &\quad + s_a\left(x_b\mathbf{i}+y_b\mathbf{j}+z_b\mathbf{k}\right) \\ &\quad + s_b\left(x_a\mathbf{i}+y_a\mathbf{j}+z_a\mathbf{k}\right) \\ &\quad + \left(y_az_b-y_bz_a\right)\mathbf{i}+\left(x_bz_a-x_az_b\right)\mathbf{j}+\left(x_ay_b-x_by_a\right)\mathbf{k}\big] \\ &= \big[s_as_b-x_ax_b-y_ay_b-z_az_b, \\ &\quad + \left(s_ax_b+s_bx_a+y_az_b-y_bz_a\right)\mathbf{i} \\ &\quad + \left(s_ay_b+s_by_a+x_bz_a-x_az_b\right)\mathbf{j} \\ &\quad + \left(s_az_b+s_bz_a+x_ay_b-x_by_a\right)\mathbf{k}\big] \\ &= \begin{bmatrix} s_a & -x_a & -y_a & -z_a \\ x_a & s_a & -z_a & y_a \\ y_a & z_a & s_a & -x_a \\ z_a & -y_a & x_a & s_a \end{bmatrix} \begin{bmatrix} s_b \\ x_b \\ y_b \\ z_b \end{bmatrix} = \mathbf{A}q_b. \end{aligned}$$

At this stage we have quaternion q_a represented by matrix **A**, and quaternion q_b represented as a column vector. Now let's reverse the scenario without altering the result by making q_b the matrix and q_a the column vector:

$$q_a q_b = \begin{bmatrix} s_b & -x_b & -y_b & -z_b \\ x_b & s_b & z_b & -y_b \\ y_b & -z_b & s_b & x_b \\ z_b & y_b & -x_b & s_b \end{bmatrix} \begin{bmatrix} s_a \\ x_a \\ y_a \\ z_a \end{bmatrix} = \mathbf{B} q_a.$$

So now we have two ways of computing $q_a q_b$ and we need a way of distinguishing between the two matrices. Let **L** be the matrix that preserves the left-to-right quaternion sequence, and **R** be the matrix that reverses the sequence to right-to-left:

$$q_a q_b = \mathbf{L}(q_a) q_b = \begin{bmatrix} s_a & -x_a & -y_a & -z_a \\ x_a & s_a & -z_a & y_a \\ y_a & z_a & s_a & -x_a \\ z_a & -y_a & x_a & s_a \end{bmatrix} \begin{bmatrix} s_b \\ x_b \\ y_b \\ z_b \end{bmatrix}$$

$$q_a q_b = \mathbf{R}(q_b) q_a = \begin{bmatrix} s_b & -x_b & -y_b & -z_b \\ x_b & s_b & z_b & -y_b \\ y_b & -z_b & s_b & x_b \\ z_b & y_b & -x_b & s_b \end{bmatrix} \begin{bmatrix} s_a \\ x_a \\ y_a \\ z_a \end{bmatrix}.$$

Remember that $\mathbf{L}(q_a) q_b = \mathbf{R}(q_b) q_a$, as this is central to understanding the next stage. Furthermore, don't be surprised if you can't follow the argument in the first reading. It took the author many hours of anguish trying to decipher the original algorithm, and this explanation has been expanded to ensure that you do not suffer the same experience!

First, let's employ the matrices **L** and **R** to rearrange the quaternion product $q_a q_c q_b$ to $q_a q_b q_c$. i.e. move q_c from the middle to the right-hand-side. We start with the quaternion product $q_a q_c q_b$ and divide it into two parts, $q_a q_c$ and q_b. We can do this because quaternion algebra is associative:

$$q_a q_c q_b = (q_a q_c) q_b.$$

We have already demonstrated above that the product $q_a q_c$ can be replaced by $\mathbf{L}(q_a) q_c$:

$$q_a q_c q_b = \mathbf{L}(q_a) q_c q_b.$$

We now have another two parts: $\mathbf{L}(q_a) q_c$ and q_b which can be reversed using **R** without disturbing the result:

$$q_a q_c q_b = \mathbf{L}(q_a) q_c q_b = \mathbf{R}(q_b) \mathbf{L}(q_a) q_c$$

which has achieved our objective to move q_c to the right-hand-side. But the most important result is that the matrices $\mathbf{R}(q_b)$ and $\mathbf{L}(q_a)$ can be multiplied together to form a single matrix, which operates on q_c.

Now let's repeat the same process to rearrange the product qpq^{-1}. The objective is to move p from the middle of q and q^{-1}, to the right-hand-side. The reason for doing this is to bring together q and q^{-1} in the form of two matrices, which can be multiplied together into a single matrix.

We start with the quaternion product qpq^{-1} and divide it into two parts, qp and q^{-1}:

$$qpq^{-1} = (qp)q^{-1}.$$

The product qp can be replaced by $\mathbf{L}(q)p$:

$$qpq^{-1} = \mathbf{L}(q)pq^{-1}.$$

We now have another two parts: $\mathbf{L}(q)p$ and q^{-1} which can be reversed using $\mathbf{R}$ without disturbing the result:

$$qpq^{-1} = \mathbf{L}(q)pq^{-1} = \mathbf{R}(q^{-1})\mathbf{L}(q)p$$

which has achieved our objective to move p to the right-hand-side.

The next step is to compute $\mathbf{L}(q)$ and $\mathbf{R}(q^{-1})$ using $q = [s,\ x\mathbf{i} + y\mathbf{j} + z\mathbf{k}]$. $\mathbf{L}(q)$ is easy as it is the same as $\mathbf{L}(q_a)$:

$$\mathbf{L}(q) = \begin{bmatrix} s & -x & -y & -z \\ x & s & -z & y \\ y & z & s & -x \\ z & -y & x & s \end{bmatrix}.$$

$\mathbf{R}(q^{-1})$ is also easy, but requires converting q_b in the original definition into q^{-1} which is effected by reversing the signs of the vector components:

$$\mathbf{R}(q^{-1}) = \begin{bmatrix} s & x & y & z \\ -x & s & -z & y \\ -y & z & s & -x \\ -z & -y & x & s \end{bmatrix}.$$

So now we can write

$$qpq^{-1} = \mathbf{R}(q^{-1})\mathbf{L}(q)p$$

$$= \begin{bmatrix} s & x & y & z \\ -x & s & -z & y \\ -y & z & s & -x \\ -z & -y & x & s \end{bmatrix} \begin{bmatrix} s & -x & -y & -z \\ x & s & -z & y \\ y & z & s & -x \\ z & -y & x & s \end{bmatrix} \begin{bmatrix} 0 \\ x_p \\ y_p \\ z_p \end{bmatrix}$$

$$= \begin{bmatrix} 1 & 0 & 0 & 0 \\ 0 & 1-2\left(y^2+z^2\right) & 2(xy-sz) & 2(xz+sy) \\ 0 & 2(xy+sz) & 1-2\left(x^2+z^2\right) & 2(yz-sx) \\ 0 & 2(xz-sy) & 2(yz+sx) & 1-2\left(x^2+y^2\right) \end{bmatrix} \begin{bmatrix} 0 \\ x_p \\ y_p \\ z_p \end{bmatrix}.$$

If we ignore the first row and column, the matrix computes $\mathbf{p}'$:

$$\mathbf{p}' = \begin{bmatrix} 1-2\left(y^2+z^2\right) & 2(xy-sz) & 2(xz+sy) \\ 2(xy+sz) & 1-2\left(x^2+z^2\right) & 2(yz-sx) \\ 2(xz-sy) & 2(yz+sx) & 1-2\left(x^2+y^2\right) \end{bmatrix} \begin{bmatrix} x_p \\ y_p \\ z_p \end{bmatrix}$$

which is identical to (8.12)!

8.4.3 Geometric Verification

Let's illustrate the action of (8.11) by rotating the point $(0, 1, 1)$, $90°$ about the y-axis, as shown in Fig. 8.6. The quaternion takes the form

$$q = \left[\cos\left(\tfrac{\theta}{2}\right),\ \sin\left(\tfrac{\theta}{2}\right)\hat{\mathbf{v}}\right]$$

which means that $\theta = 90°$ and $\hat{\mathbf{v}} = \mathbf{j}$, therefore,

$$q = \left[\cos 45°,\ \sin 45°\hat{\mathbf{j}}\right].$$

Consequently,

$$s = \tfrac{\sqrt{2}}{2},\quad x = 0,\quad y = \tfrac{\sqrt{2}}{2},\quad z = 0.$$

Substituting these values in (8.11) gives

$$\mathbf{p}' = \begin{bmatrix} 2\left(s^2+x^2\right)-1 & 2(xy-sz) & 2(xz+sy) \\ 2(xy+sz) & 2\left(s^2+y^2\right)-1 & 2(yz-sx) \\ 2(xz-sy) & 2(yz+sx) & 2\left(s^2+z^2\right)-1 \end{bmatrix} \begin{bmatrix} x_p \\ y_p \\ z_p \end{bmatrix}$$

$$\begin{bmatrix} 1 \\ 1 \\ 0 \end{bmatrix} = \begin{bmatrix} 0 & 0 & 1 \\ 0 & 1 & 0 \\ -1 & 0 & 0 \end{bmatrix} \begin{bmatrix} 0 \\ 1 \\ 1 \end{bmatrix}$$

where $(0,\ 1,\ 1)$ is rotated to $(1,\ 1,\ 0)$, which is correct.

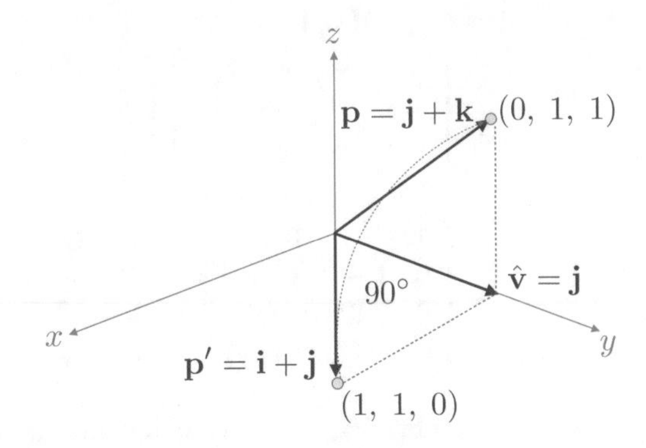

Fig. 8.6 The point $P(0, 1, 1)$ is rotated 90° to $P'(1, 1, 0)$ about the y-axis

So now we have a transform that rotates a point about an arbitrary axis intersecting the origin without the problems of gimbal lock associated with Euler transforms.

Before moving on, let's evaluate one more example. Let's perform a 180° rotation about a vector $\mathbf{v} = \mathbf{i} + \mathbf{k}$. To begin with, we will deliberately forget to convert the $\mathbf{v}$ into a unit vector, just to see what happens to the final matrix. The quaternion takes the form

$$q = \left[\cos\left(\tfrac{\theta}{2}\right),\ \sin\left(\tfrac{\theta}{2}\right)\hat{\mathbf{v}}\right]$$

but we will use $\mathbf{v}$ as specified. Therefore, with $\theta = 180°$

$$s = 0, \quad x = 1, \quad y = 0, \quad z = 1.$$

Substituting these values in (8.11) gives

$$\begin{aligned}
\mathbf{p}' &= \begin{bmatrix} 2\left(s^2 + x^2\right) - 1 & 2(xy - sz) & 2(xz + sy) \\ 2(xy + sz) & 2\left(s^2 + y^2\right) - 1 & 2(yz - sx) \\ 2(xz - sy) & 2(yz + sx) & 2\left(s^2 + z^2\right) - 1 \end{bmatrix} \begin{bmatrix} x_p \\ y_p \\ z_p \end{bmatrix} \\
&= \begin{bmatrix} 1 & 0 & 2 \\ 0 & -1 & 0 \\ 2 & 0 & 1 \end{bmatrix} \begin{bmatrix} 1 \\ 0 \\ 0 \end{bmatrix}
\end{aligned}$$

which looks nothing like a rotation matrix, and reminds us how important it is to have a unit vector to represent the axis. Let's repeat these calculations normalising the vector to $\hat{\mathbf{v}} = \frac{\sqrt{2}}{2}\mathbf{i} + \frac{\sqrt{2}}{2}\mathbf{k}$:

$$s = 0, \quad x = \tfrac{\sqrt{2}}{2}, \quad y = 0, \quad z = \tfrac{\sqrt{2}}{2}.$$

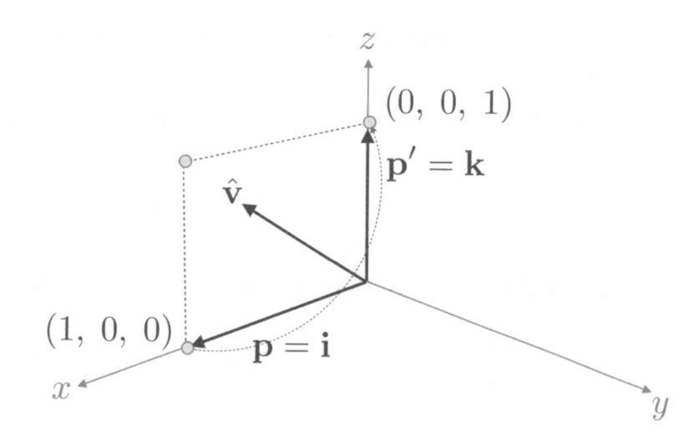

Fig. 8.7 The point $P(1,\ 0,\ 0)$ is rotated 180° to $P'(0,\ 0,\ 1)$ about $\hat{\mathbf{v}}$

Substituting these values in (8.11) gives

$$\mathbf{p}' = \begin{bmatrix} 2\left(s^2+x^2\right)-1 & 2\left(xy-sz\right) & 2\left(xz+sy\right) \\ 2\left(xy+sz\right) & 2\left(s^2+y^2\right)-1 & 2\left(yz-sx\right) \\ 2\left(xz-sy\right) & 2\left(yz+sx\right) & 2\left(s^2+z^2\right)-1 \end{bmatrix} \begin{bmatrix} x_p \\ y_p \\ z_p \end{bmatrix}$$

$$\begin{bmatrix} 0 \\ 0 \\ 1 \end{bmatrix} = \begin{bmatrix} 0 & 0 & 1 \\ 0 & -1 & 0 \\ 1 & 0 & 0 \end{bmatrix} \begin{bmatrix} 1 \\ 0 \\ 0 \end{bmatrix}$$

which not only looks like a rotation matrix, but has a determinant of 1 and rotates the point $(1,\ 0,\ 0)$ to $(0,\ 0,\ 1)$ as shown in Fig. 8.7.

8.5 Multiple Rotations

Say a vector or frame of reference is subjected to two rotations specified by q_1 followed by q_2. There is a temptation to convert both quaternions to their respective matrix and multiply the matrices together. However, this not the most efficient way of combining the rotations. It is best to accumulate the rotations as quaternions and then convert to matrix notation, if required.

To illustrate this, consider the pure quaternion p subjected to the first quaternion q_1:

$$q_1 p q_1^{-1}$$

followed by a second quaternion q_2

$$q_2 \left(q_1 p q_1^{-1}\right) q_2^{-1}$$

which can be expressed as

$$\left(q_2 q_1\right) p \left(q_2 q_1\right)^{-1}.$$

Extra quaternions can be added accordingly. Let's illustrate this with two examples.

To keep things simple, the first quaternion q_1 rotates 30° about the y-axis:

$$q_1 = \left[\cos 15^\circ,\ \sin 15^\circ \mathbf{j}\right].$$

The second quaternion q_2 rotates 60° also about the y-axis:

$$q_2 = \left[\cos 30^\circ,\ \sin 30^\circ \mathbf{j}\right].$$

Together, the two quaternions rotate 90° about the y-axis. To accumulate these rotations, we multiply them together:

$$\begin{aligned}
q_1 q_2 &= \left[\cos 15^\circ,\ \sin 15^\circ \mathbf{j}\right]\left[\cos 30^\circ,\ \sin 30^\circ \mathbf{j}\right] \\
&= \left[\cos 15^\circ \cos 30^\circ - \sin 15^\circ \sin 30^\circ,\ \cos 15^\circ \sin 30^\circ \mathbf{j} + \cos 30^\circ \sin 15^\circ \mathbf{j}\right] \\
&= \tfrac{\sqrt{2}}{2}\left[1,\ \mathbf{j}\right]
\end{aligned}$$

which is a quaternion that rotates 90° about the y-axis. Using the matrix (8.11) we have

$$\begin{aligned}
\mathbf{p}' &= \begin{bmatrix} 2\left(s^2 + x^2\right) - 1 & 2(xy - sz) & 2(xz + sy) \\ 2(xy + sz) & 2\left(s^2 + y^2\right) - 1 & 2(yz - sx) \\ 2(xz - sy) & 2(yz + sx) & 2\left(s^2 + z^2\right) - 1 \end{bmatrix}\begin{bmatrix} x_p \\ y_p \\ z_p \end{bmatrix} \\
&= \begin{bmatrix} 0 & 0 & 1 \\ 0 & 1 & 0 \\ -1 & 0 & 0 \end{bmatrix}\begin{bmatrix} x_p \\ y_p \\ z_p \end{bmatrix}
\end{aligned}$$

which rotates points about the y-axis by 90°.

For a second example, let's just evaluate the quaternions. The first quaternion q_1 rotates 90° about the x-axis, and q_2 rotates 90° about the y-axis:

$$\begin{aligned}
q_1 &= \tfrac{\sqrt{2}}{2}\left[1,\ \mathbf{i}\right] \\
q_2 &= \tfrac{\sqrt{2}}{2}\left[1,\ \mathbf{j}\right] \\
p &= \left[0,\ \mathbf{i} + \mathbf{j}\right].
\end{aligned}$$

Therefore,

$$\begin{aligned}
q_2 q_1 &= \tfrac{\sqrt{2}}{2}\left[1,\ \mathbf{i}\right]\tfrac{\sqrt{2}}{2}\left[1,\ \mathbf{j}\right] \\
&= \tfrac{1}{2}\left[1,\ \mathbf{i} + \mathbf{j} - \mathbf{k}\right] \\
(q_2 q_1)^{-1} &= \tfrac{1}{2}\left[1,\ -\mathbf{i} - \mathbf{j} + \mathbf{k}\right]
\end{aligned}$$

$$
\begin{aligned}
(q_2 q_1)\,p &= \tfrac{1}{2}\,[1,\ \mathbf{i}+\mathbf{j}-\mathbf{k}]\,[0,\ \mathbf{i}+\mathbf{j}] \\
&= \tfrac{1}{2}\,[-2,\ (\mathbf{i}+\mathbf{j})+\mathbf{i}-\mathbf{j}] \\
&= [-1,\ \mathbf{i}] \\
(q_2 q_1)\,p(q_2 q_1)^{-1} &= \tfrac{1}{2}\,[-1,\ \mathbf{i}]\,[1,\ -\mathbf{i}-\mathbf{j}+\mathbf{k}] \\
&= \tfrac{1}{2}\,[-1+1,\ \mathbf{i}+\mathbf{j}-\mathbf{k}+\mathbf{i}-\mathbf{j}-\mathbf{k}] \\
&= [0,\ \mathbf{i}-\mathbf{k}]\,.
\end{aligned}
$$

Thus the point $(1,\ 1,\ 0)$ is rotated to $(1,\ 0,\ -1)$, which is correct.

8.6 Rotating About an Off-Set Axis

Now that we have a matrix to represent a quaternion rotor, we can employ it to resolve problems such as rotating a point about an off-set axis using the same techniques associated with normal rotation transforms. For example, in Chap. 7 we used the following notation:

$$
\begin{bmatrix} x' \\ y' \\ z' \\ 1 \end{bmatrix} = \mathbf{T}_{(t_x,\ 0,\ t_z)}\mathbf{R}_{\beta,\ y}\mathbf{T}_{(-t_x,\ 0,\ -t_z)} \begin{bmatrix} x \\ y \\ z \\ 1 \end{bmatrix}
$$

to rotate a point about a fixed axis parallel with the y-axis. Therefore, by substituting the matrix qpq^{-1} for $\mathbf{R}_{\beta,\ y}$ we have:

$$
\begin{bmatrix} x' \\ y' \\ z' \\ 1 \end{bmatrix} = \mathbf{T}_{(t_x,\ 0,\ t_z)}\left(qpq^{-1}\right)\mathbf{T}_{(-t_x,\ 0,\ -t_z)} \begin{bmatrix} x \\ y \\ z \\ 1 \end{bmatrix}.
$$

Let's test this by rotating our unit cube 90° about the axis intersecting vertices 4 and 6 as shown in Fig. 8.8.

The unit-norm quaternion to achieve this is

$$
q = \left[\cos 45^\circ,\ \sin 45^\circ \mathbf{j}\right]
$$

with the pure quaternion

$$
p = [0,\ \mathbf{p}].
$$

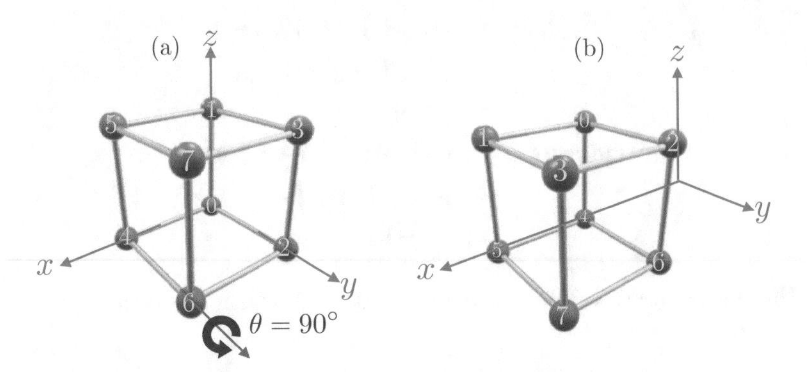

Fig. 8.8 The cube is rotated 90° about the axis intersecting vertices 4 and 6

Consequently,

$$s = \tfrac{\sqrt{2}}{2}, \quad x = 0, \quad y = \tfrac{\sqrt{2}}{2}, \quad z = 0$$

and using (8.11) in a homogeneous form we have

$$\mathbf{p}' = \begin{bmatrix} 2\left(s^2 + x^2\right) - 1 & 2(xy - sz) & 2(xz + sy) & 0 \\ 2(xy + sz) & 2\left(s^2 + y^2\right) - 1 & 2(yz - sx) & 0 \\ 2(xz - sy) & 2(yz + sx) & 2\left(s^2 + z^2\right) - 1 & 0 \\ 0 & 0 & 0 & 1 \end{bmatrix} \begin{bmatrix} x_p \\ y_p \\ z_p \\ 1 \end{bmatrix}$$

$$= \begin{bmatrix} 0 & 0 & 1 & 0 \\ 0 & 1 & 0 & 0 \\ -1 & 0 & 0 & 0 \\ 0 & 0 & 0 & 1 \end{bmatrix} \begin{bmatrix} x_p \\ y_p \\ z_p \\ 1 \end{bmatrix}.$$

The other two matrices are

$$\mathbf{T}_{(-t_x,\,0,\,0)} = \begin{bmatrix} 1 & 0 & 0 & -1 \\ 0 & 1 & 0 & 0 \\ 0 & 0 & 1 & 0 \\ 0 & 0 & 0 & 1 \end{bmatrix}$$

$$\mathbf{T}_{(t_x,\,0,\,0)} = \begin{bmatrix} 1 & 0 & 0 & 1 \\ 0 & 1 & 0 & 0 \\ 0 & 0 & 1 & 0 \\ 0 & 0 & 0 & 1 \end{bmatrix}.$$

Multiplying these three matrices together creates:

$$\mathbf{p}'\mathbf{T}_{(-t_x,\,0,\,0)} = \begin{bmatrix} 0 & 0 & 1 & 0 \\ 0 & 1 & 0 & 0 \\ -1 & 0 & 0 & 0 \\ 0 & 0 & 0 & 1 \end{bmatrix} \begin{bmatrix} 1 & 0 & 0 & -1 \\ 0 & 1 & 0 & 0 \\ 0 & 0 & 1 & 0 \\ 0 & 0 & 0 & 1 \end{bmatrix} = \begin{bmatrix} 0 & 0 & 1 & 0 \\ 0 & 1 & 0 & 0 \\ -1 & 0 & 0 & 1 \\ 0 & 0 & 0 & 1 \end{bmatrix}$$

$$\mathbf{T}_{(t_x,\,0,\,0)}\mathbf{p}'\mathbf{T}_{(-t_x,\,0,\,0)} = \begin{bmatrix} 1 & 0 & 0 & 1 \\ 0 & 1 & 0 & 0 \\ 0 & 0 & 1 & 0 \\ 0 & 0 & 0 & 1 \end{bmatrix} \begin{bmatrix} 0 & 0 & 1 & 0 \\ 0 & 1 & 0 & 0 \\ -1 & 0 & 0 & 1 \\ 0 & 0 & 0 & 1 \end{bmatrix} = \begin{bmatrix} 0 & 0 & 1 & 1 \\ 0 & 1 & 0 & 0 \\ -1 & 0 & 0 & 1 \\ 0 & 0 & 0 & 1 \end{bmatrix}$$

$$\mathbf{T}_{(t_x,\,0,\,0)}\mathbf{p}'\mathbf{T}_{(-t_x,\,0,\,0)} = \begin{bmatrix} 0 & 0 & 1 & 1 \\ 0 & 1 & 0 & 0 \\ -1 & 0 & 0 & 1 \\ 0 & 0 & 0 & 1 \end{bmatrix}. \tag{8.15}$$

Although not mathematically correct, the following statement shows the matrix (8.15) and the array of coordinates representing a unit cube, followed by the rotated cube's coordinates.

$$\begin{bmatrix} 0 & 0 & 1 & 1 \\ 0 & 1 & 0 & 0 \\ -1 & 0 & 0 & 1 \\ 0 & 0 & 0 & 1 \end{bmatrix} \begin{bmatrix} 0 & 0 & 0 & 0 & 1 & 1 & 1 & 1 \\ 0 & 0 & 1 & 1 & 0 & 0 & 1 & 1 \\ 0 & 1 & 0 & 1 & 0 & 1 & 0 & 1 \\ 1 & 1 & 1 & 1 & 1 & 1 & 1 & 1 \end{bmatrix} = \begin{bmatrix} 1 & 2 & 1 & 2 & 1 & 2 & 1 & 2 \\ 0 & 0 & 1 & 1 & 0 & 0 & 1 & 1 \\ 1 & 1 & 1 & 1 & 0 & 0 & 0 & 0 \\ 1 & 1 & 1 & 1 & 1 & 1 & 1 & 1 \end{bmatrix}.$$

These coordinates are confirmed by Fig. 8.8.

8.7 Frames of Reference

In computer graphics we often have to compute an object's coordinates relative to two frames of reference, i.e. different axial systems. For example, Fig. 8.9a shows two axial systems A_1 and A_2, that coincide with one another, and then A_2 is rotated 90° about the common z-axis. Figure 8.9b shows a unit cube which has coordinates relative to A_1 *and* A_2. As the coordinates are known for A_1, then the coordinates relative to A_2 are calculated by rotating the unit cube by an equal angle in the opposite direction associated with the frame of reference. In general, whatever transform is used to separate the two frames of reference, the opposite transform is applied to the object's coordinates.

If a quaternion is used to separate two frames of reference, then we can calculate an object's coordinates by using the inverse quaternion triple $q^{-1}pq$, which rotates in the opposite direction. To demonstrate this, consider the problem described above, and shown in Fig. 8.9. The unit-norm quaternion for such a rotation is

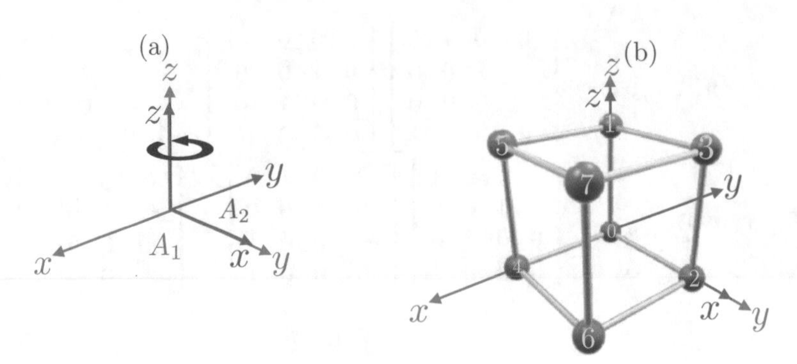

Fig. 8.9 Frame of reference A_1 is rotated 90° about the z-axis to A_2

$$
\begin{aligned}
q &= \left[\cos 45°,\ \sin 45° \mathbf{k}\right] \\
&= \left[\tfrac{\sqrt{2}}{2},\ \tfrac{\sqrt{2}}{2}\mathbf{k}\right].
\end{aligned}
$$

Consequently,

$$
s = \tfrac{\sqrt{2}}{2}, \quad x = 0, \quad y = 0, \quad z = \tfrac{\sqrt{2}}{2}.
$$

Substituting these values in (8.13) we obtain:

$$
\begin{aligned}
q^{-1}pq &= \begin{bmatrix} 2\left(s^2+x^2\right)-1 & 2(xy+sz) & 2(xz-sy) \\ 2(xy-sz) & 2\left(s^2+y^2\right)-1 & 2(yz+sx) \\ 2(xz+sy) & 2(yz-sx) & 2\left(s^2+z^2\right)-1 \end{bmatrix} \begin{bmatrix} x_p \\ y_p \\ z_p \end{bmatrix} \\
&= \begin{bmatrix} 0 & 1 & 0 \\ -1 & 0 & 0 \\ 0 & 0 & 1 \end{bmatrix} \begin{bmatrix} x_p \\ y_p \\ z_p \end{bmatrix}
\end{aligned}
$$

which, if used to process the coordinates of our unit cube, produces

$$
\begin{bmatrix} 0 & 1 & 0 \\ -1 & 0 & 0 \\ 0 & 0 & 1 \end{bmatrix} \begin{bmatrix} 0 & 0 & 0 & 0 & 1 & 1 & 1 & 1 \\ 0 & 0 & 1 & 1 & 0 & 0 & 1 & 1 \\ 0 & 1 & 0 & 1 & 0 & 1 & 0 & 1 \end{bmatrix} = \begin{bmatrix} 0 & 0 & 1 & 1 & 0 & 0 & 1 & 1 \\ 0 & 0 & 0 & 0 & -1 & -1 & -1 & -1 \\ 0 & 1 & 0 & 1 & 0 & 1 & 0 & 1 \end{bmatrix}.
$$

This scenario is shown in Fig. 8.9b.

8.8 Interpolating

Interpolation describes the process of blending one quantity into another. The process is controlled by an algorithm describing the nature of the blend function, which may be linear, quadratic, cubic, spherical, etc. If you wish to learn more about this subject,

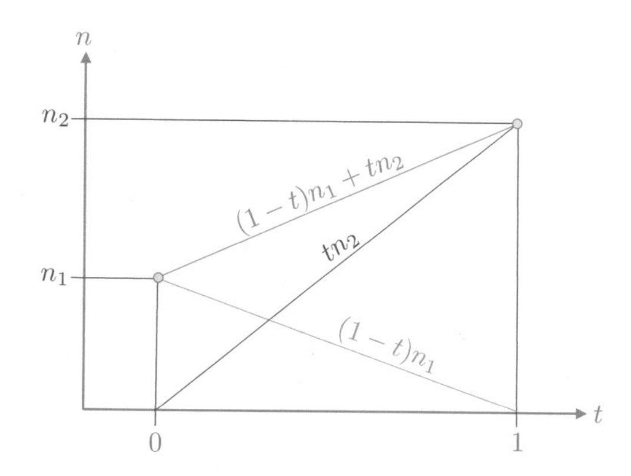

Fig. 8.10 Graphs of the functions for linear interpolation

it is covered in the author's book *Mathematics for Computer Graphics* [3]. Let's begin with a linear blend function.

8.8.1 *Linear Interpolation*

Say we have two numbers, n_1 and n_2, and a third, n_t, such that $n_1 \leq n_t \leq n_2$, where n_t is controlled by the parameter t, such that $t \in [0,\ 1]$. Then we can write:

$$n_t = (1 - t)n_1 + tn_2$$

which is a linear interpolant, and is called a *lerp*. For example, if $n_1 = 12$ and $n_2 = 20$, then the number half-way between 12 and 20 is:

$$16 = (1 - 0.5) \times 12 + 0.5 \times 20.$$

Similarly, the number a quarter-way between 12 and 20 is:

$$16 = (1 - 0.25) \times 12 + 0.25 \times 20.$$

Figure 8.10 shows graphically the relationships between $n_1,\ n_2$ and t.

8.9 Interpolating Vectors

So far we have been interpolating between a pair of numbers. Now the question arises: can we use the same interpolants for vectors? We can if we interpolate both the magnitude and direction of a vector. However, if we linearly interpolate only the x- and y-components of two vectors, the in-between vectors would neither respect

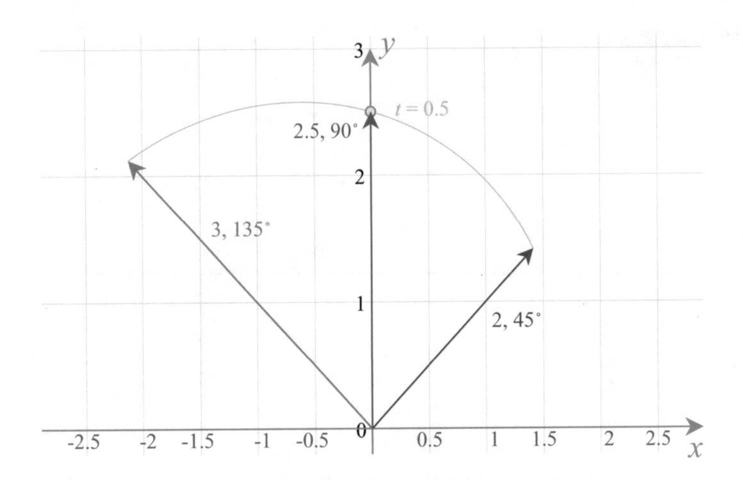

Fig. 8.11 The trace of interpolating between vectors 2, 45° and 3, 135°

their orientation or their magnitude. But if we defined two 2-D vectors as $l_1,\ \theta_1$ and $l_2,\ \theta_2$, where l is the magnitude and θ the rotated angle, then a linearly interpolated vector is given by

$$
\begin{aligned}
l &= (1-t)l_1 + tl_2 \\
\theta &= (1-t)\theta_1 + t\theta_2
\end{aligned}
$$

and the x- and y-components of the interpolated vector are:

$$
\begin{aligned}
l_x &= l\cos\theta \\
l_y &= l\sin\theta.
\end{aligned}
$$

Figure 8.11 shows the trace of interpolating between vector 2, 45° and vector 3, 135°. The half-way point, when $t = 0.5$, generates the vector 2.5, 90°. The same technique can be used with 3-D vectors using the equivalent polar notation.

We can interpolate between x- y- and z-coordinates if we respect the magnitude and orientation of the encoded vectors using the following technique. Figure 8.12 shows two unit vectors $\mathbf{v}_1$ and $\mathbf{v}_2$ separated by an angle θ. The interpolated vector $\mathbf{v}$ is defined as a proportion of $\mathbf{v}_1$ and a proportion of $\mathbf{v}_2$:

$$
\mathbf{v} = a\mathbf{v}_1 + b\mathbf{v}_2.
$$

Let's define the values of a and b such that they are a function of the separating angle θ. Vector $\mathbf{v}$ is $t\theta$ from $\mathbf{v}_1$ and $(1-t)\theta$ from $\mathbf{v}_2$, and it is evident from Fig. 8.12 that using the sine rule

$$
\frac{a}{\sin[(1-t)\theta]} = \frac{b}{\sin(t\theta)} \tag{8.16}
$$

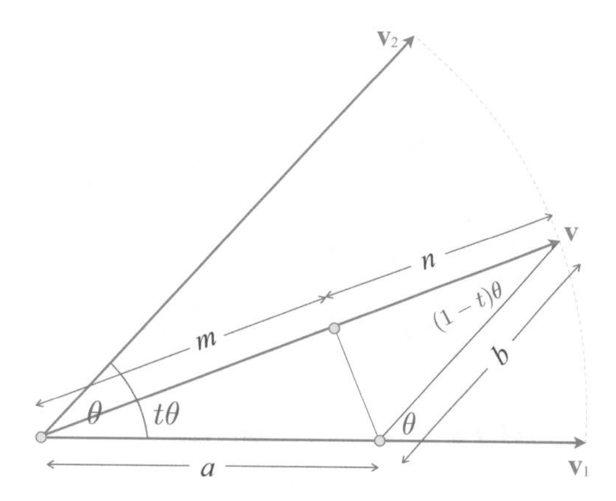

Fig. 8.12 Vector $\mathbf{v}$ is derived from part a of of $\mathbf{v}_1$ and part b of $\mathbf{v}_2$

and furthermore:

$$
\begin{aligned}
m &= a\cos(t\theta) \\
n &= b\cos[(1-t)\theta]
\end{aligned}
$$

where

$$
m + n = 1. \tag{8.17}
$$

From (8.16)

$$
b = \frac{a\sin(t\theta)}{\sin[(1-t)\theta]}
$$

and from (8.17) we get

$$
a\cos(t\theta) + \frac{a\sin(t\theta)\cos[(1-t)\theta]}{\sin[(1-t)\theta]} = 1.
$$

Solving for a we find

$$
\begin{aligned}
a &= \frac{\sin[(1-t)\theta]}{\sin\theta} \\
b &= \frac{\sin(t\theta)}{\sin\theta}.
\end{aligned}
$$

Therefore, the final interpolant is

$$
\mathbf{v} = \frac{\sin[(1-t)\theta]}{\sin\theta}\mathbf{v}_1 + \frac{\sin(t\theta)}{\sin\theta}\mathbf{v}_2. \tag{8.18}
$$

Equation (8.18) is called a *slerp*, which is short for spherical, linear interpolant.

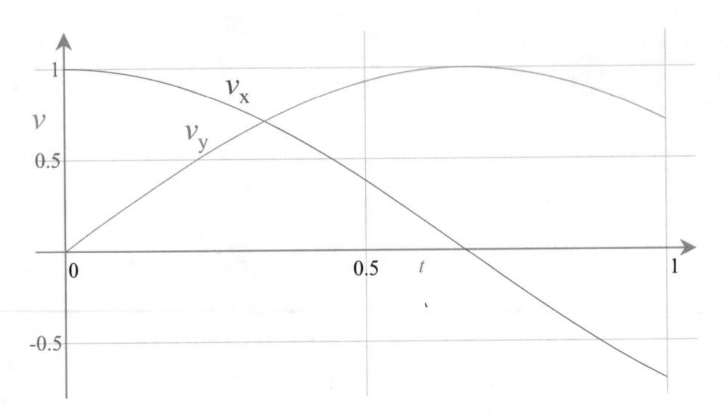

Fig. 8.13 Curves of v_x and v_y using (8.18)

To see how (8.18) operates, let's consider a simple exercise of interpolating between two unit vectors $[1 \quad 0]^T$ and $[-1/\sqrt{2} \quad 1/\sqrt{2}]^T$. The angle between the vectors θ is 135°. Equation (8.18) is used to interpolate the x- and the y-components individually:

$$v_x = \frac{\sin[(1-t)135°]}{\sin 135°} \times (1) + \frac{\sin(t135°)}{\sin 135°} \times \left(-\frac{1}{\sqrt{2}}\right)$$
$$v_y = \frac{\sin[(1-t)135°]}{\sin 135°} \times (0) + \frac{\sin(t135°)}{\sin 135°} \times \left(\frac{1}{\sqrt{2}}\right).$$

Figure 8.13 shows the interpolating curves and Fig. 8.14 shows a trace of the interpolated vectors.

Two observations to note with (8.18):

- The angle θ is the angle between the two vectors, which, if not known, can be computed using the dot product.
- Secondly, the range of θ is given by $0 \le \theta \le 180°$, but when $\theta = 180°$ the denominator collapses to zero.

So far, we have only considered unit vectors. Now let's see how the interpolant reacts to vectors of different magnitudes. As a test, we can input the following vectors to (8.18):

$$\mathbf{v}_1 = [2 \quad 0]^T, \quad \text{and} \quad \mathbf{v}_2 = [0 \quad 1]^T.$$

The separating angle $\theta = 90°$, and the result is shown in Fig. 8.15. Note how the initial length of $\mathbf{v}_1$ reduces from 2 to 1 over 90°. It is left to the reader to examine other combinations of vectors. There is one more application for this interpolant, and that is with quaternions.

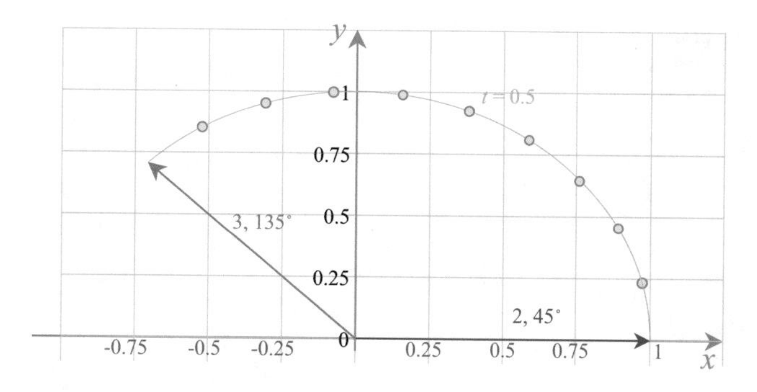

Fig. 8.14 A trace of the interpolated vectors $[1 \quad 0]^{\mathrm{T}}$ and $\left[-1/\sqrt{2} \quad 1/\sqrt{2}\right]^{\mathrm{T}}$

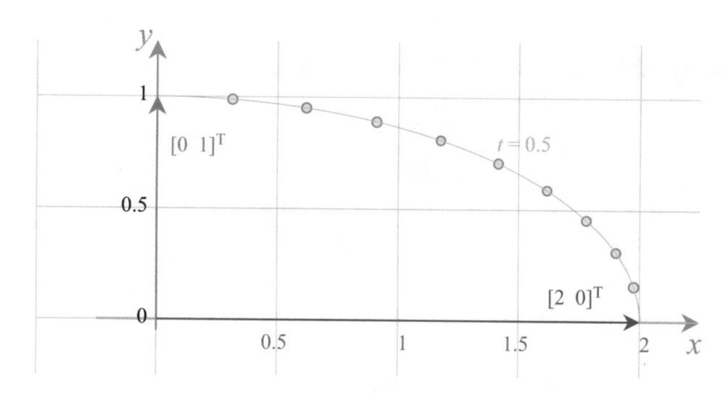

Fig. 8.15 Interpolating between the vectors $[2 \quad 0]^{\mathrm{T}}$ and $[0 \quad 1]^{\mathrm{T}}$

8.10 Interpolating Quaternions

Like vectors, quaternions can be interpolated to compute an in-between quaternion. However, whereas two interpolated vectors results in a third vector that is readily visualised, two interpolated quaternions results in a third quaternion that acts as a rotor, and is not immediately visualised.

The spherical interpolant for quaternions is

$$q = \frac{\sin[(1-t)\theta]}{\sin\theta} q_1 + \frac{\sin(t\theta)}{\sin\theta} q_2. \tag{8.19}$$

So, given

$$\begin{aligned} q_1 &= [s_1, \ x_1\mathbf{i} + y_1\mathbf{j} + z_1\mathbf{k}] \\ q_2 &= [s_2, \ x_2\mathbf{i} + y_2\mathbf{j} + z_2\mathbf{k}] \end{aligned}$$

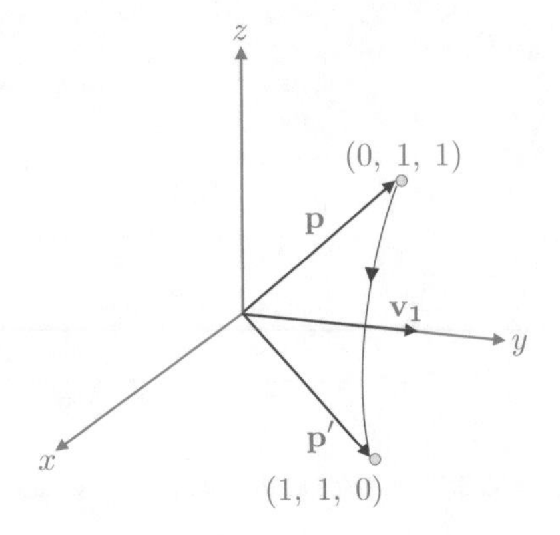

Fig. 8.16 The point $(0,\ 1,\ 1)$ is rotated 90° about the vector $\mathbf{v}_1$ to $(1,\ 1,\ 0)$

θ is obtained by taking the 4-D dot product of q_1 and q_2:

$$\begin{aligned}\cos\theta &= \frac{q_1 q_2}{|q_1||q_2|}\\ &= \frac{s_1 s_2 + x_1 x_2 + y_1 y_2 + z_1 z_2}{|q_1||q_2|}\end{aligned}$$

and if we are working with unit-norm quaternions, then

$$\cos\theta = s_1 s_2 + x_1 x_2 + y_1 y_2 + z_1 z_2. \tag{8.20}$$

Let's use (8.19) in a scenario with two simple unit-norm quaternions.

Figure 8.16 shows one such scenario where the point $(0,\ 1,\ 1)$ is rotated 90° about $\mathbf{v}_1$, the axis of q_1. Figure 8.17 shows another scenario where the same point $(0,\ 1,\ 1)$ is rotated 90° about $\mathbf{v}_2$, the axis of q_2. The quaternions are

$$\begin{aligned}q_1 &= \left[\cos 45^\circ,\ \sin 45^\circ\mathbf{j}\right] = \left[\tfrac{\sqrt{2}}{2},\ \tfrac{\sqrt{2}}{2}\mathbf{j}\right]\\ q_2 &= \left[\cos 45^\circ,\ \sin 45^\circ\mathbf{i}\right] = \left[\tfrac{\sqrt{2}}{2},\ \tfrac{\sqrt{2}}{2}\mathbf{i}\right].\end{aligned}$$

Therefore, using (8.20)

$$\begin{aligned}\cos\theta &= \tfrac{\sqrt{2}}{2}\tfrac{\sqrt{2}}{2} = 0.5\\ \theta &= 60^\circ.\end{aligned}$$

Before proceeding, let's compute the matrices for the two quaternion products. For q_1:

$$q_1 = [\cos 45^\circ,\ \sin 45^\circ\mathbf{j}]$$

$$s = \tfrac{\sqrt{2}}{2},\quad x = 0,\quad y = \tfrac{\sqrt{2}}{2},\quad z = 0$$

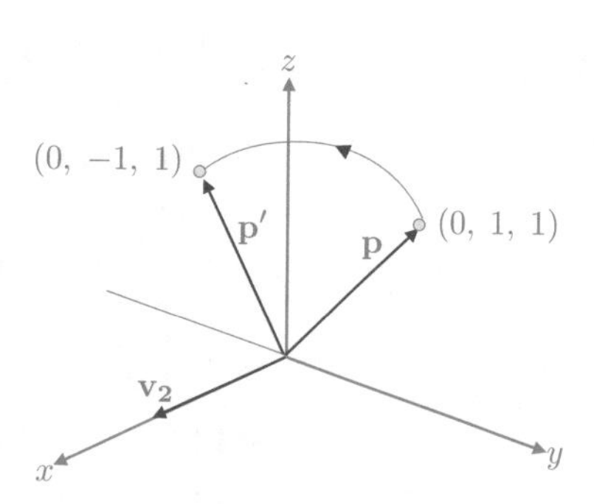

Fig. 8.17 The point $(0,\ 1,\ 1)$ is rotated 90° about the vector $\mathbf{v}_2$ to $(0,\ -1,\ 1)$

which when substituted in (8.11) gives

$$\mathbf{p}_1' = \begin{bmatrix} 2(s^2+x^2)-1 & 2(xy-sz) & 2(xz+sy) \\ 2(xy+sz) & 2(s^2+y^2)-1 & 2(yz-sx) \\ 2(xz-sy) & 2(yz+sx) & 2(s^2+z^2)-1 \end{bmatrix} \begin{bmatrix} x_p \\ y_p \\ z_p \end{bmatrix}$$

$$= \begin{bmatrix} 0 & 0 & 1 \\ 0 & 1 & 0 \\ -1 & 0 & 0 \end{bmatrix} \begin{bmatrix} x_p \\ y_p \\ z_p \end{bmatrix}. \qquad (8.21)$$

Substituting the coordinates $(0,\ 1,\ 1)$ in (8.21) gives $(1,\ 1,\ 0)$.

For q_2:

$$q_2 = [\cos 45°,\ \sin 45°\mathbf{i}]$$

$$s = \tfrac{\sqrt{2}}{2}, \quad x = \tfrac{\sqrt{2}}{2}, \quad y = 0, \quad z = 0$$

which when substituted in (8.11) gives

$$\mathbf{p}_2' = \begin{bmatrix} 2(s^2+x^2)-1 & 2(xy-sz) & 2(xz+sy) \\ 2(xy+sz) & 2(s^2+y^2)-1 & 2(yz-sx) \\ 2(xz-sy) & 2(yz+sx) & 2(s^2+z^2)-1 \end{bmatrix} \begin{bmatrix} x_p \\ y_p \\ z_p \end{bmatrix}$$

$$= \begin{bmatrix} 1 & 0 & 0 \\ 0 & 0 & -1 \\ 0 & 1 & 0 \end{bmatrix} \begin{bmatrix} x_p \\ y_p \\ z_p \end{bmatrix}. \qquad (8.22)$$

Substituting the coordinates $(0,\ 1,\ 1)$ in (8.22) gives $(0,\ -1,\ 1)$.

Now let's compute (8.19) for five values of $t = 0,\ 0.25,\ 0.5,\ 0.75,\ 1$. Note that the original quaternions have a unit norm, which implies that any interpolated quaternion must also have a unit norm.

When $t = 0$, (8.19) will give q_1, which means $(0,\ 1,\ 1)$ is rotated to $(1,\ 1,\ 0)$.

When $t = 0.25$, (8.19) is:

$$
\begin{aligned}
q &= \frac{\sin[(1-t)\theta]}{\sin\theta}q_1 + \frac{\sin(t\theta)}{\sin\theta}q_2 \\
&= \frac{\sin\left(\frac{180^\circ}{4}\right)}{\sin 60^\circ}\left[\tfrac{\sqrt{2}}{2},\ \tfrac{\sqrt{2}}{2}\mathbf{j}\right] + \frac{\sin\left(\frac{60^\circ}{4}\right)}{\sin 60^\circ}\left[\tfrac{\sqrt{2}}{2},\ \tfrac{\sqrt{2}}{2}\mathbf{i}\right] \\
&= \frac{\sin 45^\circ}{\sin 60^\circ}\left[\tfrac{\sqrt{2}}{2},\ \tfrac{\sqrt{2}}{2}\mathbf{j}\right] + \frac{\sin 15^\circ}{\sin 60^\circ}\left[\tfrac{\sqrt{2}}{2},\ \tfrac{\sqrt{2}}{2}\mathbf{i}\right] \\
&\approx \tfrac{\sqrt{2}}{\sqrt{3}}\left[\tfrac{\sqrt{2}}{2},\ \tfrac{\sqrt{2}}{2}\mathbf{j}\right] + \tfrac{0.5176}{\sqrt{3}}\left[\tfrac{\sqrt{2}}{2},\ \tfrac{\sqrt{2}}{2}\mathbf{i}\right] \\
&\approx [0.5774,\ 0.5774\mathbf{j}] + [0, 2113,\ 0.2113\mathbf{i}] \\
&\approx [0.7887,\ 0.2113\mathbf{i} + 0.5774\mathbf{j}] \\
|q| &= 1
\end{aligned}
$$

where

$$
s \approx 0.7887, \quad x \approx 0.2113, \quad y \approx 0.5774, \quad z = 0
$$

which when substituted in (8.11) gives

$$
\mathbf{p}' \approx \begin{bmatrix} 0.3333 & 0.2441 & 0.9109 \\ 0.2441 & 0.9109 & -0.3333 \\ -0.9109 & 0.3333 & 0.2441 \end{bmatrix} \begin{bmatrix} x_p \\ y_p \\ z_p \end{bmatrix}. \tag{8.23}
$$

Substituting the coordinates $(0,\ 1,\ 1)$ in (8.23) gives $(1.155,\ 0.577,\ 0.577)$.

When $t = 0.5$, (8.19) computes a mid-way position for an interpolated quaternion, with its vector at 45° between the x- and y-axes, as shown in Fig. 8.18.

$$
\begin{aligned}
q &= \frac{\sin[(1-t)\theta]}{\sin\theta}q_1 + \frac{\sin t\theta}{\sin\theta}q_2 \\
&= \frac{\sin\left(\frac{60^\circ}{2}\right)}{\sin 60^\circ}\left[\tfrac{\sqrt{2}}{2},\ \tfrac{\sqrt{2}}{2}\mathbf{j}\right] + \frac{\sin\left(\frac{60^\circ}{2}\right)}{\sin 60^\circ}\left[\tfrac{\sqrt{2}}{2},\ \tfrac{\sqrt{2}}{2}\mathbf{i}\right] \\
&= \tfrac{1}{\sqrt{3}}\left[\tfrac{\sqrt{2}}{2},\ \tfrac{\sqrt{2}}{2}\mathbf{j}\right] + \tfrac{1}{\sqrt{3}}\left[\tfrac{\sqrt{2}}{2},\ \tfrac{\sqrt{2}}{2}\mathbf{i}\right] \\
&= \left[\tfrac{\sqrt{2}}{\sqrt{3}},\ \tfrac{1}{\sqrt{6}}\mathbf{i} + \tfrac{1}{\sqrt{6}}\mathbf{j}\right]
\end{aligned}
$$

where

$$
s = \tfrac{\sqrt{2}}{\sqrt{3}}, \quad x = \tfrac{1}{\sqrt{6}}, \quad y = \tfrac{1}{\sqrt{6}}, \quad z = 0
$$

which when substituted in (8.11) gives

$$
\mathbf{p}' = \begin{bmatrix} \frac{2}{3} & \frac{1}{3} & \frac{2}{3} \\ \frac{1}{3} & \frac{2}{3} & -\frac{2}{3} \\ -\frac{2}{3} & \frac{2}{3} & \frac{1}{3} \end{bmatrix} \begin{bmatrix} x_p \\ y_p \\ z_p \end{bmatrix}. \tag{8.24}
$$

Substituting the coordinates $(0,\ 1,\ 1)$ in (8.24) gives $(1,\ 0,\ 1)$.

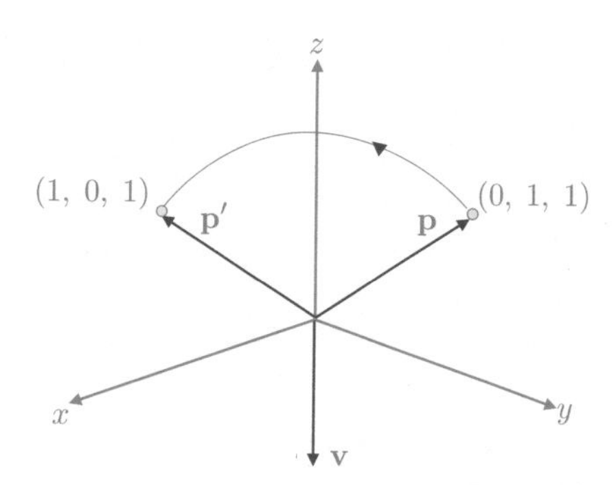

Fig. 8.18 The point (0, 1, 1) is rotated 90° about the vector **v** to (1, 0, 1)

When $t = 0.75$, (8.19) is:

$$
\begin{aligned}
q &= \frac{\sin[(1-t)\theta]}{\sin\theta} q_1 + \frac{\sin(t\theta)}{\sin\theta} q_2 \\
&= \frac{\sin\left(\frac{60^\circ}{4}\right)}{\sin 60^\circ}\left[\tfrac{\sqrt{2}}{2},\ \tfrac{\sqrt{2}}{2}\mathbf{j}\right] + \frac{\sin\left(\frac{180^\circ}{4}\right)}{\sin 60^\circ}\left[\tfrac{\sqrt{2}}{2},\ \tfrac{\sqrt{2}}{2}\mathbf{i}\right] \\
&= \frac{\sin 15^\circ}{\sin 60^\circ}\left[\tfrac{\sqrt{2}}{2},\ \tfrac{\sqrt{2}}{2}\mathbf{j}\right] + \frac{\sin 45^\circ}{\sin 60^\circ}\left[\tfrac{\sqrt{2}}{2},\ \tfrac{\sqrt{2}}{2}\mathbf{i}\right] \\
&\approx \tfrac{0.5176}{\sqrt{3}}\left[\tfrac{\sqrt{2}}{2},\ \tfrac{\sqrt{2}}{2}\mathbf{j}\right] + \tfrac{\sqrt{2}}{\sqrt{3}}\left[\tfrac{\sqrt{2}}{2},\ \tfrac{\sqrt{2}}{2}\mathbf{i}\right] \\
&\approx [0,2113,\ 0,2113\mathbf{j}] + [0.5774,\ 0.5774\mathbf{i}] \\
&\approx [0.7887,\ 0.5774\mathbf{i} + 0,2113\mathbf{j}] \\
|q| &= 1
\end{aligned}
$$

where

$$s \approx 0.7887, \quad x \approx 0.5774, \quad y \approx 0,2113, \quad z = 0$$

which when substituted in (8.11) gives

$$
\mathbf{p}' \approx \begin{bmatrix} 0.9109 & 0.2441 & 0.3333 \\ 0.2441 & 0.3333 & -0.9109 \\ -0.3333 & 0.9109 & 0.2441 \end{bmatrix} \begin{bmatrix} x_p \\ y_p \\ z_p \end{bmatrix}. \tag{8.25}
$$

Substituting the coordinates (0, 1, 1) in (8.25) gives (0.577, −0.577, 1.155).

The interpolated points are shown in Table 8.1, and Fig. 8.19 shows a sketch of these points.

Table 8.1 The points created by interpolating quaternions q_1 and q_2

t	x	y	z
0	1	1	0
0.25	1.155	0.577	0.577
0.5	1	0	1
0.75	0.577	−0.577	1.155
1	0	−1	1

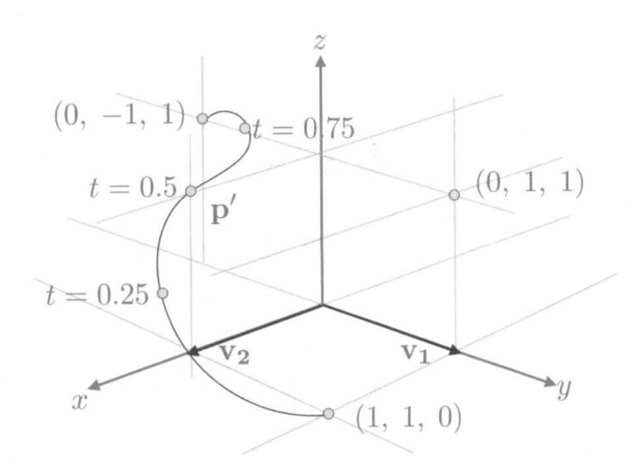

Fig. 8.19 A sketch of the path generated by interpolating two quaternions

Let us remind ourselves how to interpret Fig. 8.19. There are two quaternions, q_1 and q_2, that rotate 90° about the y-axis and x-axis, respectively. These quaternions possess unit vectors $\mathbf{v}_1$ and $\mathbf{v}_2$ directed along the y-axis and x-axis, respectively. Figure 8.19 shows an example point $(0,\ 1,\ 1)$ which is rotated to $(1,\ 1,\ 0)$ by q_1, and $(0,\ -1,\ 1)$ by q_2. When the spherical interpolant (8.11), is used to create an inbetween quaternion q, which acts on the point $(0,\ 1,\ 1)$, it traces out the path shown in Fig. 8.19.

One of the reasons for using a spherical interpolant is that it linearly interpolates the angle between the two unit-norm quaternions, which creates a constant-angular velocity between them. However, one of the problems with visualising quaternions is that they reside in a four-dimensional space and create a hyper-sphere with a radius equal to the quaternion's norm. With our 3-D brains, this is difficult to visualise. Nevertheless, we can convince ourselves into thinking we see what is going on with a simple sketch, as shown in Fig. 8.20, where we see part of the hyper-sphere and two quaternions q_1 and q_2. In this example, the angle ϕ is a constant angle between two values of the interpolant t. The spherical interpolant also ensures that the norm of the interpolated quaternion remains constant at unity, preventing any unwanted scaling.

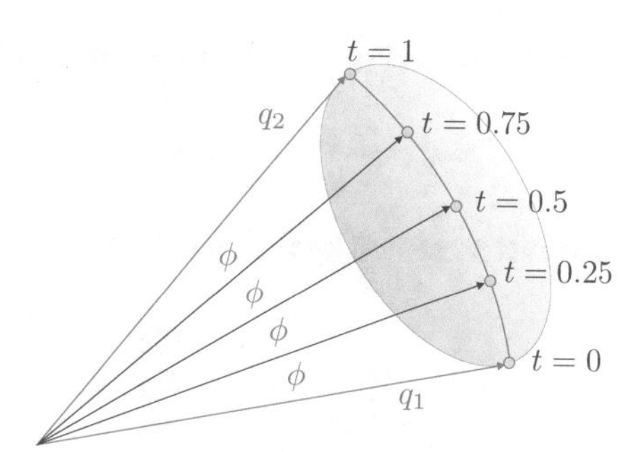

Fig. 8.20 Spherical interpolation between q_1 and q_2

8.11 Converting a Rotation Matrix to a Quaternion

Very often one has a 3-D rotation matrix which would be nice to see as a quaternion. So let's see how this can be realised. The matrix transform equivalent to qpq^{-1} is

$$qpq^{-1} = \begin{bmatrix} 2\left(s^2+x^2\right)-1 & 2(xy-sz) & 2(xz+sy) \\ 2(xy+sz) & 2\left(s^2+y^2\right)-1 & 2(yz-sx) \\ 2(xz-sy) & 2(yz+sx) & 2\left(s^2+z^2\right)-1 \end{bmatrix} \begin{bmatrix} x_p \\ y_p \\ z_p \end{bmatrix} \tag{8.26}$$

$$= \begin{bmatrix} a_{11} & a_{12} & a_{13} \\ a_{21} & a_{22} & a_{23} \\ a_{31} & a_{32} & a_{33} \end{bmatrix} \begin{bmatrix} x_p \\ y_p \\ z_p \end{bmatrix}. \tag{8.27}$$

Inspection of (8.26) and (8.27) shows that by combining various elements we can isolate the terms of a quaternion $s,\ x,\ y,\ z$. For example, by adding the diagonal terms of (8.27): $a_{11}+a_{22}+a_{33}$, we obtain

$$\begin{aligned} a_{11}+a_{22}+a_{33} &= \left[2\left(s^2+x^2\right)-1\right]+\left[2\left(s^2+y^2\right)-1\right]+\left[2\left(s^2+z^2\right)-1\right] \\ &= 6s^2+2\left(x^2+y^2+z^2\right)-3 \\ &= 4s^2-1 \end{aligned}$$

therefore,

$$s = \tfrac{1}{2}\sqrt{1+a_{11}+a_{22}+a_{33}}.$$

To isolate $x,\ y,\ z$ we employ:

$$\begin{aligned} x &= \frac{1}{4s}\left(a_{32}-a_{23}\right) \\ y &= \frac{1}{4s}\left(a_{13}-a_{31}\right) \\ z &= \frac{1}{4s}\left(a_{21}-a_{12}\right). \end{aligned}$$

We can test this inversion using the interpolated quaternion:

$$q = \left[\frac{\sqrt{2}}{\sqrt{3}}, \frac{1}{\sqrt{6}}\mathbf{i} + \frac{1}{\sqrt{6}}\mathbf{j} + 0\mathbf{k}\right]$$

and is represented by the matrix (8.24):

$$\begin{bmatrix} a_{11} & a_{12} & a_{13} \\ a_{21} & a_{22} & a_{23} \\ a_{31} & a_{32} & a_{33} \end{bmatrix} = \begin{bmatrix} \frac{2}{3} & \frac{1}{3} & \frac{2}{3} \\ \frac{1}{3} & \frac{2}{3} & -\frac{2}{3} \\ -\frac{2}{3} & \frac{2}{3} & \frac{1}{3} \end{bmatrix}$$

$$\begin{aligned} s &= \tfrac{1}{2}\sqrt{1 + a_{11} + a_{22} + a_{33}} &&= \tfrac{1}{2}\sqrt{1 + \tfrac{2}{3} + \tfrac{2}{3} + \tfrac{1}{3}} &&= \tfrac{\sqrt{2}}{\sqrt{3}} \\ x &= \tfrac{1}{4s}(a_{32} - a_{23}) &&= \tfrac{\sqrt{3}}{4\sqrt{2}}\left(\tfrac{2}{3} + \tfrac{2}{3}\right) &&= \tfrac{1}{\sqrt{6}} \\ y &= \tfrac{1}{4s}(a_{13} - a_{31}) &&= \tfrac{\sqrt{3}}{4\sqrt{2}}\left(\tfrac{2}{3} + \tfrac{2}{3}\right) &&= \tfrac{1}{\sqrt{6}} \\ z &= \tfrac{1}{4s}(a_{21} - a_{12}) &&= \tfrac{\sqrt{3}}{4\sqrt{2}}\left(\tfrac{1}{3} - \tfrac{1}{3}\right) &&= 0 \end{aligned}$$

which agree with the original values.

Observe that this technique depends on s being non-zero, in which case, other combinations are available, which also rely upon non-zero elements:

$$\begin{aligned} x &= \tfrac{1}{2}\sqrt{1 + a_{11} - a_{22} - a_{33}} &&= \tfrac{1}{2}\sqrt{1 + \tfrac{2}{3} - \tfrac{2}{3} - \tfrac{1}{3}} &&= \tfrac{1}{\sqrt{6}} \\ y &= \tfrac{1}{4x}(a_{12} + a_{21}) &&= \tfrac{\sqrt{6}}{4}\left(\tfrac{1}{3} + \tfrac{1}{3}\right) &&= \tfrac{1}{\sqrt{6}} \\ z &= \tfrac{1}{4x}(a_{13} + a_{31}) &&= \tfrac{\sqrt{6}}{4}\left(\tfrac{2}{3} - \tfrac{2}{3}\right) &&= 0 \\ s &= \tfrac{1}{4x}(a_{32} - a_{23}) &&= \tfrac{\sqrt{6}}{4}\left(\tfrac{2}{3} + \tfrac{2}{3}\right) &&= \tfrac{\sqrt{2}}{\sqrt{3}}. \end{aligned}$$

$$\begin{aligned} y &= \tfrac{1}{2}\sqrt{1 - a_{11} + a_{22} - a_{33}} &&= \tfrac{1}{2}\sqrt{1 - \tfrac{2}{3} + \tfrac{2}{3} - \tfrac{1}{3}} &&= \tfrac{1}{\sqrt{6}} \\ x &= \tfrac{1}{4y}(a_{12} + a_{21}) &&= \tfrac{\sqrt{6}}{4}\left(\tfrac{1}{3} + \tfrac{1}{3}\right) &&= \tfrac{1}{\sqrt{6}} \\ z &= \tfrac{1}{4y}(a_{23} + a_{32}) &&z = \tfrac{1}{4y}\left(-\tfrac{2}{3} + \tfrac{2}{3}\right) &&= 0 \\ s &= \tfrac{1}{4y}(a_{13} - a_{31}) &&z = \tfrac{1}{4y}\left(\tfrac{2}{3} + \tfrac{2}{3}\right) &&= \tfrac{\sqrt{2}}{\sqrt{3}}. \end{aligned}$$

$$\begin{aligned} z &= \tfrac{1}{2}\sqrt{1 - a_{11} - a_{22} + a_{33}} &&= \tfrac{1}{2}\sqrt{1 - \tfrac{2}{3} - \tfrac{2}{3} + \tfrac{1}{3}} &&= 0 \\ x &= \tfrac{1}{4z}(a_{13} + a_{31}) \\ y &= \tfrac{1}{4z}(a_{23} + a_{32}) \\ s &= \tfrac{1}{4z}(a_{21} - a_{12}). \end{aligned}$$

Note that the last set of equations create a divide by zero, which is inadmissible.

8.12 Euler Angles to Quaternion

In Chap. 7 we discovered that the rotation transforms $\mathbf{R}_{\alpha,\,x}$, $\mathbf{R}_{\beta,\,y}$ and $\mathbf{R}_{\gamma,\,z}$ can be combined to create twelve triple combinations to represent a composite rotation. Now let's see how such a transform is represented by a quaternion.

To demonstrate the technique we must choose one of the twelve combinations, then the same technique can be used to convert other combinations. For example, let's choose the sequence $\mathbf{R}_{\gamma,\,z}\mathbf{R}_{\beta,\,y}\mathbf{R}_{\alpha,\,x}$ where the equivalent quaternions are

$$\begin{aligned}
q_x &= \left[\cos\left(\tfrac{\alpha}{2}\right),\ \sin\left(\tfrac{\alpha}{2}\right)\mathbf{i}\right]\\
q_y &= \left[\cos\left(\tfrac{\beta}{2}\right),\ \sin\left(\tfrac{\beta}{2}\right)\mathbf{j}\right]\\
q_z &= \left[\cos\left(\tfrac{\gamma}{2}\right),\ \sin\left(\tfrac{\gamma}{2}\right)\mathbf{k}\right]
\end{aligned}$$

and

$$q = q_z q_y q_x. \tag{8.28}$$

Expanding (8.28):

$$\begin{aligned}
q &= \left[\cos\left(\tfrac{\gamma}{2}\right),\ \sin\left(\tfrac{\gamma}{2}\right)\mathbf{k}\right]\left[\cos\left(\tfrac{\beta}{2}\right),\ \sin\left(\tfrac{\beta}{2}\right)\mathbf{j}\right]\left[\cos\left(\tfrac{\alpha}{2}\right),\ \sin\left(\tfrac{\alpha}{2}\right)\mathbf{i}\right]\\
&= \Big[\cos\left(\tfrac{\gamma}{2}\right)\cos\left(\tfrac{\beta}{2}\right),\\
&\quad \cos\left(\tfrac{\gamma}{2}\right)\sin\left(\tfrac{\beta}{2}\right)\mathbf{j}+\cos\left(\tfrac{\beta}{2}\right)\sin\left(\tfrac{\gamma}{2}\right)\mathbf{k}-\sin\left(\tfrac{\gamma}{2}\right)\sin\left(\tfrac{\beta}{2}\right)\mathbf{i}\Big]\left[\cos\left(\tfrac{\alpha}{2}\right),\ \sin\left(\tfrac{\alpha}{2}\right)\mathbf{i}\right]\\
&= \Big[\cos\left(\tfrac{\gamma}{2}\right)\cos\left(\tfrac{\beta}{2}\right)\cos\left(\tfrac{\alpha}{2}\right)+\sin\left(\tfrac{\gamma}{2}\right)\sin\left(\tfrac{\beta}{2}\right)\sin\left(\tfrac{\alpha}{2}\right),\\
&\quad \cos\left(\tfrac{\gamma}{2}\right)\cos\left(\tfrac{\beta}{2}\right)\sin\left(\tfrac{\alpha}{2}\right)\mathbf{i}+\cos\left(\tfrac{\alpha}{2}\right)\cos\left(\tfrac{\gamma}{2}\right)\sin\left(\tfrac{\beta}{2}\right)\mathbf{j}+\cos\left(\tfrac{\alpha}{2}\right)\cos\left(\tfrac{\beta}{2}\right)\sin\left(\tfrac{\gamma}{2}\right)\mathbf{k}\\
&\quad -\cos\left(\tfrac{\alpha}{2}\right)\sin\left(\tfrac{\gamma}{2}\right)\sin\left(\tfrac{\beta}{2}\right)\mathbf{i}-\cos\left(\tfrac{\gamma}{2}\right)\sin\left(\tfrac{\beta}{2}\right)\sin\left(\tfrac{\alpha}{2}\right)\mathbf{k}+\cos\left(\tfrac{\beta}{2}\right)\sin\left(\tfrac{\gamma}{2}\right)\sin\left(\tfrac{\alpha}{2}\right)\mathbf{j}\Big]\\
&= \Big[\cos\left(\tfrac{\gamma}{2}\right)\cos\left(\tfrac{\beta}{2}\right)\cos\left(\tfrac{\alpha}{2}\right)+\sin\left(\tfrac{\gamma}{2}\right)\sin\left(\tfrac{\beta}{2}\right)\sin\left(\tfrac{\alpha}{2}\right),\\
&\quad \left(\cos\left(\tfrac{\gamma}{2}\right)\cos\left(\tfrac{\beta}{2}\right)\sin\left(\tfrac{\alpha}{2}\right)-\cos\left(\tfrac{\alpha}{2}\right)\sin\left(\tfrac{\gamma}{2}\right)\sin\left(\tfrac{\beta}{2}\right)\right)\mathbf{i}\\
&\quad \left(\cos\left(\tfrac{\alpha}{2}\right)\cos\left(\tfrac{\gamma}{2}\right)\sin\left(\tfrac{\beta}{2}\right)+\cos\left(\tfrac{\beta}{2}\right)\sin\left(\tfrac{\gamma}{2}\right)\sin\left(\tfrac{\alpha}{2}\right)\right)\mathbf{j}\\
&\quad \left(\cos\left(\tfrac{\alpha}{2}\right)\cos\left(\tfrac{\beta}{2}\right)\sin\left(\tfrac{\gamma}{2}\right)-\cos\left(\tfrac{\gamma}{2}\right)\sin\left(\tfrac{\beta}{2}\right)\sin\left(\tfrac{\alpha}{2}\right)\right)\mathbf{k}\Big].
\end{aligned}$$

Now let's place the angles in a consistent sequence:

$$
\begin{aligned}
s &= \cos\left(\tfrac{\gamma}{2}\right)\cos\left(\tfrac{\beta}{2}\right)\cos\left(\tfrac{\alpha}{2}\right) + \sin\left(\tfrac{\gamma}{2}\right)\sin\left(\tfrac{\beta}{2}\right)\sin\left(\tfrac{\alpha}{2}\right)\\
x_q &= \cos\left(\tfrac{\gamma}{2}\right)\cos\left(\tfrac{\beta}{2}\right)\sin\left(\tfrac{\alpha}{2}\right) - \sin\left(\tfrac{\gamma}{2}\right)\sin\left(\tfrac{\beta}{2}\right)\cos\left(\tfrac{\alpha}{2}\right)\\
y_q &= \cos\left(\tfrac{\gamma}{2}\right)\sin\left(\tfrac{\beta}{2}\right)\cos\left(\tfrac{\alpha}{2}\right) + \sin\left(\tfrac{\gamma}{2}\right)\cos\left(\tfrac{\beta}{2}\right)\sin\left(\tfrac{\alpha}{2}\right)\\
z_q &= \sin\left(\tfrac{\gamma}{2}\right)\cos\left(\tfrac{\beta}{2}\right)\cos\left(\tfrac{\alpha}{2}\right) - \cos\left(\tfrac{\gamma}{2}\right)\sin\left(\tfrac{\beta}{2}\right)\sin\left(\tfrac{\alpha}{2}\right)
\end{aligned}
$$

where

$$
q = \left[s,\ x_q\mathbf{i} + y_q\mathbf{j} + z_q\mathbf{k}\right]. \tag{8.29}
$$

Let's test (8.29). We start with the three rotation transforms:

$$
\begin{aligned}
\mathbf{R}_{\alpha,\,x} &= \begin{bmatrix} 1 & 0 & 0\\ 0 & \cos\alpha & -\sin\alpha\\ 0 & \sin\alpha & \cos\alpha \end{bmatrix}\\
\mathbf{R}_{\beta,\,y} &= \begin{bmatrix} \cos\beta & 0 & \sin\beta\\ 0 & 1 & 0\\ -\sin\beta & 0 & \cos\beta \end{bmatrix}\\
\mathbf{R}_{\gamma,\,z} &= \begin{bmatrix} \cos\gamma & -\sin\gamma & 0\\ \sin\gamma & \cos\gamma & 0\\ 0 & 0 & 1 \end{bmatrix}.
\end{aligned}
$$

Then

$$
\mathbf{R}_{\gamma,\,z}\mathbf{R}_{\beta,\,y}\mathbf{R}_{\alpha,\,x} =
$$

$$
\begin{bmatrix}
\cos\gamma\cos\beta & -\sin\gamma\cos\alpha + \cos\gamma\sin\beta\sin\alpha & \sin\gamma\sin\alpha + \cos\gamma\sin\beta\cos\alpha\\
\sin\gamma\cos\beta & \cos\gamma\cos\alpha + \sin\gamma\sin\beta\sin\alpha & -\cos\gamma\sin\alpha + \sin\gamma\sin\beta\cos\alpha\\
-\sin\beta & \cos\beta\sin\alpha & \cos\beta\cos\alpha
\end{bmatrix}.
$$

Let's make $\alpha = \beta = \gamma = 90°$, then

$$
\mathbf{R}_{90°,\,z}\mathbf{R}_{90°,\,y}\mathbf{R}_{90°,\,x} = \begin{bmatrix} 0 & 0 & 1\\ 0 & 1 & 0\\ -1 & 0 & 0 \end{bmatrix}
$$

which rotates points 90° about the y-axis:

$$
\begin{bmatrix} 1\\ 1\\ 0 \end{bmatrix} = \begin{bmatrix} 0 & 0 & 1\\ 0 & 1 & 0\\ -1 & 0 & 0 \end{bmatrix}\begin{bmatrix} 0\\ 1\\ 1 \end{bmatrix}.
$$

Now let's evaluate (8.29):

$$
\begin{aligned}
s &= \cos\left(\tfrac{\gamma}{2}\right)\cos\left(\tfrac{\beta}{2}\right)\cos\left(\tfrac{\alpha}{2}\right)+\sin\left(\tfrac{\gamma}{2}\right)\sin\left(\tfrac{\beta}{2}\right)\sin\left(\tfrac{\alpha}{2}\right)\\
&= \tfrac{\sqrt{2}}{2}\tfrac{\sqrt{2}}{2}\tfrac{\sqrt{2}}{2}+\tfrac{\sqrt{2}}{2}\tfrac{\sqrt{2}}{2}\tfrac{\sqrt{2}}{2}\\
&= \tfrac{\sqrt{2}}{2}\\
x_q &= \cos\left(\tfrac{\gamma}{2}\right)\cos\left(\tfrac{\beta}{2}\right)\sin\left(\tfrac{\alpha}{2}\right)-\sin\left(\tfrac{\gamma}{2}\right)\sin\left(\tfrac{\beta}{2}\right)\cos\left(\tfrac{\alpha}{2}\right)\\
&= 0\\
y_q &= \cos\left(\tfrac{\gamma}{2}\right)\sin\left(\tfrac{\beta}{2}\right)\cos\left(\tfrac{\alpha}{2}\right)+\sin\left(\tfrac{\gamma}{2}\right)\cos\left(\tfrac{\beta}{2}\right)\sin\left(\tfrac{\alpha}{2}\right)\\
&= \tfrac{\sqrt{2}}{2}\tfrac{\sqrt{2}}{2}\tfrac{\sqrt{2}}{2}+\tfrac{\sqrt{2}}{2}\tfrac{\sqrt{2}}{2}\tfrac{\sqrt{2}}{2}\\
&= \tfrac{\sqrt{2}}{2}\\
z_q &= \sin\left(\tfrac{\gamma}{2}\right)\cos\left(\tfrac{\beta}{2}\right)\cos\left(\tfrac{\alpha}{2}\right)-\cos\left(\tfrac{\gamma}{2}\right)\sin\left(\tfrac{\beta}{2}\right)\sin\left(\tfrac{\alpha}{2}\right)\\
&= 0
\end{aligned}
$$

and

$$
q = \left[\tfrac{\sqrt{2}}{2},\ \tfrac{\sqrt{2}}{2}\mathbf{j}\right]
$$

which is a quaternion that rotates points 90° about the y-axis.

8.13 Summary

This chapter has been the focal point of the book where unit-norm quaternions are used to rotate a vector about a quaternion's vector. It would have been useful if this could have been achieved by the simple product qp, like complex numbers. But as we saw, this only works when the quaternion is orthogonal to the vector. The product qpq^{-1}—discovered by Hamilton and Cayley—works for all orientations between a quaternion and a vector. It is also relatively easy to compute. We also saw that the product can be represented as a matrix, which can be integrated with other matrices.

Perhaps one of the most interesting features of quaternions that has emerged in this chapter, is that their imaginary qualities are not required in any calculations, because they are embedded within the algebra.

The spherical interpolant provides a clever way to dynamically change a quaternion's axis and angle of rotation, but can be difficult to visualise as an animated sequence without access to a real-time display system.

The reverse product $q^{-1}pq$ reverses the angle of rotation, and is equivalent to changing the sign of the rotation angle in qpq^{-1}. Consequently, it can be used to rotate a frame of reference in the same direction as qpq^{-1}.

8.13.1 Summary of Definitions

Rotating a vector by a quaternion

$$\begin{aligned}
q &= [s,\ \mathbf{v}] \\
s^2 + \|\mathbf{v}\|^2 &= 1 \\
p &= [0,\ \mathbf{p}] \\
qpq^{-1} &= \left[0,\ 2(\mathbf{v}\cdot\mathbf{p})\mathbf{v} + \left(2s^2 - 1\right)\mathbf{p} + 2s\mathbf{v}\times\mathbf{p}\right]. \\
q &= \left[\cos\left(\tfrac{\theta}{2}\right),\ \sin\left(\tfrac{\theta}{2}\right)\hat{\mathbf{v}}\right] \\
p &= [0,\ \mathbf{p}] \\
qpq^{-1} &= \left[0,\ (1-\cos\theta)(\hat{\mathbf{v}}\cdot\mathbf{p})\hat{\mathbf{v}} + \cos\theta\mathbf{p} + \sin\theta\hat{\mathbf{v}}\times\mathbf{p}\right].
\end{aligned}$$

Rotating a frame by a quaternion

$$q^{-1}pq = \left[0,\ (1-\cos\theta)(\hat{\mathbf{v}}\cdot\mathbf{p})\hat{\mathbf{v}} + \cos\theta\mathbf{p} - \sin\theta\hat{\mathbf{v}}\times\mathbf{p}\right].$$

Matrix for rotating a vector by a quaternion

$$\mathbf{p}' = \begin{bmatrix} 1-2\left(y^2+z^2\right) & 2(xy-sz) & 2(xz+sy) \\ 2(xy+sz) & 1-2\left(x^2+z^2\right) & 2(yz-sx) \\ 2(xz-sy) & 2(yz+sx) & 1-2\left(x^2+y^2\right) \end{bmatrix} \begin{bmatrix} x_p \\ y_p \\ z_p \end{bmatrix}.$$

Matrix for rotating a frame by a quaternion

$$\mathbf{p}' = \begin{bmatrix} 1-2\left(y^2+z^2\right) & 2(xy+sz) & 2(xz-sy) \\ 2(xy-sz) & 1-2\left(x^2+z^2\right) & 2(yz+sx) \\ 2(xz+sy) & 2(yz-sx) & 1-2\left(x^2+y^2\right) \end{bmatrix} \begin{bmatrix} x_p \\ y_p \\ z_p \end{bmatrix}.$$

Matrix for a quaternion product

$$q_1q_2 = \mathbf{L}(q_1)q_2 = \begin{bmatrix} s_1 & -x_1 & -y_1 & -z_1 \\ x_1 & s_1 & -z_1 & y_1 \\ y_1 & z_1 & s_1 & -x_1 \\ z_1 & -y_1 & x_1 & s_1 \end{bmatrix} \begin{bmatrix} s_2 \\ x_2 \\ y_2 \\ z_2 \end{bmatrix}$$

$$q_1q_2 = \mathbf{R}(q_2)q_1 = \begin{bmatrix} s_2 & -x_2 & -y_2 & -z_2 \\ x_2 & s_2 & z_2 & -y_2 \\ y_2 & -z_2 & s_2 & x_2 \\ z_2 & y_2 & -x_2 & s_2 \end{bmatrix} \begin{bmatrix} s_1 \\ x_1 \\ y_1 \\ z_1 \end{bmatrix}.$$

Interpolating two quaternions using a slerp

$$q = \frac{\sin[(1-t)\theta]}{\sin\theta} q_1 + \frac{\sin(t\theta)}{\sin\theta} q_2$$

where

$$\begin{aligned}\cos\theta &= \frac{q_1 \cdot q_2}{|q_1||q_2|} \\ &= \frac{s_1 s_2 + x_1 x_2 + y_1 y_2 + z_1 z_2}{|q_1||q_2|}.\end{aligned}$$

Quaternion from a rotation matrix

$$\begin{aligned}s &= \tfrac{1}{2}\sqrt{1 + a_{11} + a_{22} + a_{33}} \\ x &= \frac{1}{4s}(a_{32} - a_{23}) \\ y &= \frac{1}{4s}(a_{13} - a_{31}) \\ z &= \frac{1}{4s}(a_{21} - a_{12})\end{aligned}$$

$$\begin{aligned}x &= \tfrac{1}{2}\sqrt{1 + a_{11} - a_{22} - a_{33}} \\ y &= \frac{1}{4x}(a_{12} + a_{21}) \\ z &= \frac{1}{4x}(a_{13} + a_{31}) \\ s &= \frac{1}{4x}(a_{32} - a_{23})\end{aligned}$$

$$\begin{aligned}y &= \tfrac{1}{2}\sqrt{1 - a_{11} + a_{22} - a_{33}} \\ x &= \frac{1}{4y}(a_{12} + a_{21}) \\ z &= \frac{1}{4y}(a_{23} + a_{32}) \\ s &= \frac{1}{4y}(a_{13} - a_{31})\end{aligned}$$

$$
\begin{aligned}
z &= \tfrac{1}{2}\sqrt{1 - a_{11} - a_{22} + a_{33}} \\
x &= \frac{1}{4z}(a_{13} + a_{31}) \\
y &= \frac{1}{4z}(a_{23} + a_{32}) \\
s &= \frac{1}{4z}(a_{21} - a_{12}).
\end{aligned}
$$

Euler angles to quaternion
Using the transform $\mathbf{R}_{\gamma,\,z}\mathbf{R}_{\beta,\,y}\mathbf{R}_{\alpha,\,x}$:

$$
\begin{aligned}
s &= \cos\left(\tfrac{\gamma}{2}\right)\cos\left(\tfrac{\beta}{2}\right)\cos\left(\tfrac{\alpha}{2}\right) + \sin\left(\tfrac{\gamma}{2}\right)\sin\left(\tfrac{\beta}{2}\right)\sin\left(\tfrac{\alpha}{2}\right) \\
x_q &= \cos\left(\tfrac{\gamma}{2}\right)\cos\left(\tfrac{\beta}{2}\right)\sin\left(\tfrac{\alpha}{2}\right) - \sin\left(\tfrac{\gamma}{2}\right)\sin\left(\tfrac{\beta}{2}\right)\cos\left(\tfrac{\alpha}{2}\right) \\
y_q &= \cos\left(\tfrac{\gamma}{2}\right)\sin\left(\tfrac{\beta}{2}\right)\cos\left(\tfrac{\alpha}{2}\right) + \sin\left(\tfrac{\gamma}{2}\right)\cos\left(\tfrac{\beta}{2}\right)\sin\left(\tfrac{\alpha}{2}\right) \\
z_q &= \sin\left(\tfrac{\gamma}{2}\right)\cos\left(\tfrac{\beta}{2}\right)\cos\left(\tfrac{\alpha}{2}\right) - \cos\left(\tfrac{\gamma}{2}\right)\sin\left(\tfrac{\beta}{2}\right)\sin\left(\tfrac{\alpha}{2}\right)
\end{aligned}
$$

where

$$
q = \left[s,\ x_q\mathbf{i} + y_q\mathbf{j} + z_q\mathbf{k}\right].
$$

8.14 Worked Examples

Here are some further worked examples that employ the ideas described above.

8.14.1 Special Case Quaternion

Use qp to rotate $p = [0,\ \mathbf{j}]$ 90° about the x-axis.
For this to work q must be orthogonal to p:

$$
\begin{aligned}
q &= [\cos\theta,\ \sin\theta\mathbf{i}] \\
&= [0,\ \mathbf{i}]
\end{aligned}
$$

and

$$
\begin{aligned}
p' &= qp \\
&= [0,\ \mathbf{i}][0,\ \mathbf{j}] \\
&= [0,\ \mathbf{k}].
\end{aligned}
$$

8.14.2 Rotating a Vector Using a Quaternion

Use qpq^{-1} to rotate $p = [0,\ \mathbf{j}]$ 90° about the x-axis.

For this to work:

$$\begin{aligned} q &= \left[\cos\left(\tfrac{\theta}{2}\right),\ \sin\left(\tfrac{\theta}{2}\right)\mathbf{i}\right] \\ &= \left[\tfrac{\sqrt{2}}{2},\ \tfrac{\sqrt{2}}{2}\mathbf{i}\right] \\ &= \tfrac{\sqrt{2}}{2}[1,\ \mathbf{i}] \end{aligned}$$

and

$$\begin{aligned} p' &= qpq^{-1} \\ &= \tfrac{\sqrt{2}}{2}\tfrac{\sqrt{2}}{2}[1,\ \mathbf{i}]\,[0,\ \mathbf{j}]\,[1,\ -\mathbf{i}] \\ &= \tfrac{1}{2}\,[0,\ \mathbf{j}+\mathbf{k}]\,[1,\ -\mathbf{i}] \\ &= \tfrac{1}{2}\,[0,\ (\mathbf{j}+\mathbf{k})-\mathbf{j}+\mathbf{k}] \\ &= [0,\ \mathbf{k}]\,. \end{aligned}$$

8.14.3 Evaluate qpq^{-1}

Evaluate qpq^{-1} for $p = [0,\ \mathbf{p}]$ and $q = \left[\cos\left(\frac{\theta}{2}\right),\ \sin\left(\frac{\theta}{2}\right)\mathbf{v}\right]$, where $\theta = 360°$.

$$\begin{aligned} q &= [-1,\ \mathbf{0}] \\ qpq^{-1} &= [-1,\ \mathbf{0}]\,[0,\ \mathbf{p}]\,[-1,\ \mathbf{0}] \\ &= \left[0,\ -\mathbf{p}\right][-1,\ \mathbf{0}] \\ &= \left[0,\ \mathbf{p}\right] \end{aligned}$$

which confirms that the vector remains unmoved, as expected.

8.14.4 Evaluate qpq^{-1} Using a Matrix

Compute the matrix for $q = \left[\frac{1}{2},\ \frac{\sqrt{3}}{2}\mathbf{k}\right]$.

From q:

$$s = \tfrac{1}{2},\quad x = 0,\quad y = 0,\quad z = \tfrac{\sqrt{3}}{2}$$

$$\mathbf{p}' = \begin{bmatrix} 2\left(s^2 + x^2\right) - 1 & 2(xy - sz) & 2(xz + sy) \\ 2(xy + sz) & 2\left(s^2 + y^2\right) - 1 & 2(yz - sx) \\ 2(xz - sy) & 2(yz + sx) & 2\left(s^2 + z^2\right) - 1 \end{bmatrix} \begin{bmatrix} x_p \\ y_p \\ z_p \end{bmatrix}$$

$$= \begin{bmatrix} -\frac{1}{2} & -\frac{\sqrt{3}}{2} & 0 \\ \frac{\sqrt{3}}{2} & -\frac{1}{2} & 0 \\ 0 & 0 & 1 \end{bmatrix} \begin{bmatrix} x_p \\ y_p \\ z_p \end{bmatrix}.$$

If we plug in the point $(1,\ 0,\ 0)$, it is rotated about the z-axis by 120°:

$$\begin{bmatrix} -\frac{1}{2} \\ \frac{\sqrt{3}}{2} \\ 1 \end{bmatrix} = \begin{bmatrix} -\frac{1}{2} & -\frac{\sqrt{3}}{2} & 0 \\ \frac{\sqrt{3}}{2} & -\frac{1}{2} & 0 \\ 0 & 0 & 1 \end{bmatrix} \begin{bmatrix} 1 \\ 0 \\ 0 \end{bmatrix}.$$

8.14.5 Slerp Interpolation

Find the half-way quaternion between $q_1 = \left[\cos\left(\frac{\alpha}{2}\right),\ \sin\left(\frac{\alpha}{2}\right)\mathbf{k}\right]$ and $q_2 = \left[\cos\left(\frac{\alpha}{2}\right),\ \sin\left(\frac{\alpha}{2}\right)\mathbf{i}\right]$ when $\alpha = 90°$. Show that it is a unit-norm quaternion, and find its angle of rotation.

The angle between q_1 and q_2 is θ where,

$$\begin{aligned} \cos\theta &= \frac{s_1 s_2 + x_1 x_2 + y_1 y_2 + z_1 z_2}{|q_1||q_2|} \\ &= \cos^2\left(\frac{\alpha}{2}\right) = \cos^2 45° \\ &= 0.5 \\ \theta &= 60°. \end{aligned}$$

Using

$$\begin{aligned} q &= \frac{\sin[(1-t)\theta]}{\sin\theta} q_1 + \frac{\sin[t\theta]}{\sin\theta} q_2 \\ &= \frac{\sin 30°}{\sin 60°}\left[\cos 45°,\ \sin 45°\mathbf{k}\right] + \frac{\sin 30°}{\sin 60°}\left[\cos 45°,\ \sin 45°\mathbf{i}\right] \\ &= \frac{1}{\sqrt{3}}\left[\frac{\sqrt{2}}{2},\ \frac{\sqrt{2}}{2}\mathbf{k}\right] + \frac{1}{\sqrt{3}}\left[\frac{\sqrt{2}}{2},\ \frac{\sqrt{2}}{2}\mathbf{i}\right] \\ &= \left[\frac{\sqrt{2}}{\sqrt{3}},\ \frac{\sqrt{2}}{2\sqrt{3}}\mathbf{i} + \frac{\sqrt{2}}{2\sqrt{3}}\mathbf{k}\right] \\ &= \left[\frac{2}{\sqrt{6}},\ \frac{1}{\sqrt{6}}\mathbf{i} + \frac{1}{\sqrt{6}}\mathbf{k}\right]. \end{aligned}$$

The norm of q is

$$
\begin{aligned}
|q| &= \left(\tfrac{2}{\sqrt{6}}\right)^2 + \left(\tfrac{1}{\sqrt{6}}\right)^2 + \left(\tfrac{1}{\sqrt{6}}\right)^2 \\
&= \tfrac{2}{3} + \tfrac{1}{6} + \tfrac{1}{6} \\
&= 1.
\end{aligned}
$$

Therefore, $\cos\left(\frac{\alpha}{2}\right) = \frac{\sqrt{2}}{\sqrt{3}}$ and $\sin\left(\frac{\alpha}{2}\right) = \frac{1}{\sqrt{3}}$, and $\alpha \approx 70.5°$.

8.14.6 Rotation Matrix into a Quaternion

Convert the matrix $\mathbf{M}$ into a quaternion and identify its function.

$$
\mathbf{M} = \begin{bmatrix} 0 & 0 & 1 \\ 0 & 1 & 0 \\ -1 & 0 & 0 \end{bmatrix}.
$$

Therefore,

$$
\begin{aligned}
s &= \tfrac{1}{2}\sqrt{1 + a_{11} + a_{22} + a_{33}} \\
&= \tfrac{1}{2}\sqrt{1 + 0 + 1 + 0} \quad = \tfrac{\sqrt{2}}{2} \\
x &= \frac{1}{4s}\left(a_{32} - a_{23}\right) \\
&= \tfrac{\sqrt{2}}{4}\,(0 + 0) \quad = 0 \\
y &= \frac{1}{4s}\left(a_{13} - a_{31}\right) \\
&= \tfrac{\sqrt{2}}{4}\,(1 + 1) \quad = \tfrac{\sqrt{2}}{2} \\
z &= \frac{1}{4s}\left(a_{21} - a_{12}\right) \\
&= \tfrac{\sqrt{2}}{4}\,(0 + 0) \quad = 0
\end{aligned}
$$

which is the quaternion $\left[\frac{\sqrt{2}}{2}, \frac{\sqrt{2}}{2}\mathbf{j}\right]$ and is a rotation of 90° about the y-axis.

References

1. Cayley, A.: The Collected Mathematical Papers, vol. I, p. 586, note 20 (1848)
2. Altmann, S.L.: Rotations, Quaternions and Double Groups, p. 16. Dover Publications (1986). ISBN-13: 978-0-486-44518-2
3. Vince, J.A.: Mathematics for Computer Graphics, 5th edn. Springer, Berlin. ISBN 978-1-4471-7334-2

Chapter 9
Conclusion

If you have reached this chapter, having read the previous eight chapters, then there is a good chance that you have understood what a quaternion is, and how it is used to rotate vectors about an arbitrary axis. I deliberately played down the four-dimensional side of quaternions, as this feature is not relevant to understanding what they are, and how they are manipulated at an introductory level. You should now be in a position to code up the operations and discover what benefits quaternions bring to the stage of rotations. You should also be in a position to tackle more advanced texts and discover other applications.

Very rarely, has any mathematician invented something that has taken the world completely by surprise. For as we have seen with the invention of quaternions, Gauss had played around with quadruples, but was too nervous to tell anyone. Similarly, Grassmann had also been working on his own vector algebra and had written up his ideas in two books, but his style of writing was too obscure, even for mathematicians to understand! Rodrigues has been described as a 'recreational' mathematician, and perhaps just for the fun of it, decided to analyse the algebra of rotations. In so doing, he discovered a half-angle solution identical to that of the quaternion product, three years ahead of Hamilton. But in the end, it was left to Hamilton to successfully generalise complex numbers to a higher dimension. It took him over a decade to discover the final solution, and in spite of being a genius, he was unaware that a triple could not be the answer. Fortunately, his tenacity and mathematical brilliance shone through and won the day.

Although Hamilton thought that quaternions would become an important mathematical tool for a wide range of scientific applications, they were ignored in favour of the vector algebra described by Gibbs. Hamilton must have been disappointed that quaternion algebra did not become the preferred vectorial system, but he should have been extremely proud to have been responsible for the event that gave us today's vector algebra.

J. Vince, *Quaternions for Computer Graphics*,
https://doi.org/10.1007/978-1-4471-7509-4_9

It is interesting that the computer age, and especially the subject of computer graphics, has provided a useful application for quaternions. Perhaps we can at last forget about Euler rotations and work with a mathematical tool that is intuitive, easy to use and efficient—quaternions!

Index

J. Vince, *Quaternions for Computer Graphics*,
https://doi.org/10.1007/978-1-4471-7509-4

Printed in the United States
by Baker & Taylor Publisher Services